Handbook of
INTEGRATED-CIRCUIT
OPERATIONAL AMPLIFIERS

Handbook of
INTEGRATED-CIRCUIT
OPERATIONAL AMPLIFIERS

George B. Rutkowski, P.E.

Gould, Inc.
Advanced Development Division
Cleveland, Ohio

Electronic Technology Institute
Cleveland, Ohio

Prentice-Hall, Inc.
Englewood Cliffs, New Jersey

Library of Congress Cataloging in Publication Data

RUTKOWSKI, GEORGE B.
 Handbook of integrated-circuit operational amplifiers.

 1. Operational amplifiers.   2. Integrated
circuits.   I. Title.
TK7871.58 O6R87   621.381'73'5   74-13332
ISBN 0-13-378703-6

© 1975 by Prentice-Hall, Inc.,
Englewood Cliffs, N.J.

10   9   8   7

Printed in the United States of America.

PRENTICE-HALL INTERNATIONAL, INC., *London*
PRENTICE-HALL OF AUSTRALIA PTY. LTD., *Sydney*
PRENTICE-HALL OF CANADA, LTD., *Toronto*
PRENTICE-HALL OF INDIA PRIVATE LIMITED, *New Delhi*
PRENTICE-HALL OF JAPAN, INC., *Tokyo*

CONTENTS

PREFACE

Chapter 1   DIFFERENTIAL AND OPERATIONAL
            AMPLIFIER CIRCUITS   1

   1-1   Transistor review   1
   1-2   The differential circuit   5
   1-3   Current-source biased amplifiers   12
   1-4   Multi-stage differential circuits   13
   1-5   Level shifting with intermediate stage   16
   1-6   The output stage and complete operational amplifier   17

Chapter 2   CHARACTERISTICS OF OP AMPS
            AND THEIR POWER SUPPLY REQUIREMENTS   21

   2-1   Open-loop voltage gain A_{VOL} 22
   2-2   Output offset voltage V_{oo} 24
   2-3   Input resistance R_i 24
   2-4   Output resistance R_o 25
   2-5   Bandwidth BW 26
   2-6   Response time   26
   2-7   Power supply requirements   27

Chapter 3   THE OP AMP
            WITH AND WITHOUT FEEDBACK   39

   3-1   Open-loop considerations   39
   3-2   Feedback and the inverting amplifier   44
   3-3   The noninverting amplifier   50
   3-4   The voltage follower   54

Chapter 4   OFFSET CONSIDERATIONS   64

   4-1   Input offset voltage V_{io} 64
   4-2   Input bias current   68
   4-3   Input offset current   72
   4-4   Combined effects of offset voltage and offset current   74

Chapter 5   COMMON MODE VOLTAGES
            AND DIFFERENTIAL MODE AMPLIFIERS   80

    5-1   The differential mode Op Amp circuit   81
    5-2   Common-mode rejection ratio *CMRR*   83
    5-3   Maximum common mode input voltages   87
    5-4   Op Amp instrumentation circuits   89

Chapter 6   OP AMP BEHAVIOR
            AT HIGHER FREQUENCIES   100

    6-1   Gain and phase shift vs frequency   100
    6-2   Bode diagrams   102
    6-3   External frequency compensation   104
    6-4   Compensated operational amplifiers   108
    6-5   Slew rate   110
    6-6   Output swing vs frequency   115
    6-7   Noise   117
    6-8   Equivalent input noise   118

Chapter 7   PRACTICAL CONSIDERATIONS   129

    7-1   Offset voltage vs power supply voltage   129
    7-2   Offset voltage vs temperature   133
    7-3   Other temperature-sensitive parameters   138
    7-4   Programmable Op Amps   140
    7-5   Varactor- and chopper-stabilized Op Amps   145
    7-6   Channel separation   146
    7-7   Cleaning PC boards and guarding input terminals   149
    7-8   Protecting techniques   150

Chapter 8   LINEAR APPLICATIONS
            OF OP AMPS   155

    8-1   Op Amps as ac amplifiers   155
    8-2   Op Amp ac amplifier with single power supply   159
    8-3   Summing and averaging circuits   161
    8-4   The Op Amp as an integrator   166
    8-5   The Op Amp as a differentiator   169
    8-6   Op Amps in analog computers   172
    8-7   Op Amps in voltage regulators   177
    8-8   Active filters   185

Chapter **9** NONLINEAR APPLICATIONS
                OF OP AMPS   *195*

   9-1 Voltage limiters   *195*
   9-2 The zero-crossing detector   *199*
   9-3 Op Amps as comparators   *201*
   9-4 An absolute value output circuit   *202*
   9-5 The Op Amp as a small-signal diode   *205*
   9-6 The Op Amp as a sample-and-hold circuit   *206*

Chapter **10** OP AMP SIGNAL GENERATORS   *210*

   10-1 The square-wave generator   *210*
   10-2 The triangular-wave generator   *213*
   10-3 The sawtooth generator   *215*
   10-4 The twin-T oscillator   *216*
   10-5 The Wien bridge oscillator   *217*
   10-6 Variable-frequency signal generators   *218*

Chapter **11** DIGITAL APPLICATIONS
                OF OP AMPS   *222*

   11-1 The Schmitt trigger   *222*
   11-2 The monostable (one-shot) multivibrator   *224*
   11-3 Digital-to-analog (D/A) converters   *226*
   11-4 Analog-to-digital (A/D) converters   *230*

Chapter **12** THE CURRENT-DIFFERENCING
                AMPLIFIER   *234*

   12-1 Basic active section of the CD amplifier   *234*
   12-2 Feedback with the CD amplifier   *237*
   12-3 The inverting current-differencing amplifier   *240*
   12-4 The noninverting current-differencing amplifier   *242*
   12-5 The CD amplifier in differential mode   *243*
   12-6 The CD amplifier as a comparator   *245*
   12-7 The CD amplifier as a square-wave generator   *246*
   12-8 The CD amplifier as an AND gate   *247*
   12-9 The CD amplifier as an OR gate   *248*
   12-10 Collection of current differencing (Norton)
                 amplifier circuits   *249*

GLOSSARY OF TERMS   *257*

Appendix **F1**   SELECTION GUIDE FOR GENERAL
                   PURPOSE OPERATIONAL AMPLIFIERS   *263*

Appendix **F2**   SELECTION GUIDE FOR HIGH
                   ACCURACY OPERATIONAL AMPLIFIERS   *267*

Appendix **SI**   COLLECTION OF OP AMP TYPES
                   AND THEIR PARAMETERS   *270*

Appendix **SII**   SPECIFICATIONS OF THE 709 OP AMP   *272*

Appendix **SIII** SPECIFICATIONS OF THE 741 OP AMP   *276*

Appendix **SIV**   SPECIFICATIONS OF THE 777 OP AMP   *279*

Appendix **A1**   DERIVATION OF EQUATION (4-4)   *283*

Appendix **A2**   DERIVATION OF EQUATION (4-7)   *285*

Appendix **A3**   SELECTION OF COUPLING CAPACITORS   *287*

COLLECTION OF OPERATIONAL AMPLIFIER CIRCUITS   *288*

PACKAGE TYPES   *307*

ANSWERS TO ODD-NUMBERED PROBLEMS   *308*

INDEX   *311*

PREFACE

The operational amplifier (Op Amp) was developed to perform mathematical operations within analog computers. The first Op Amps were large, bulky vacuum tube types. As transistors replaced vacuum tubes, Op Amps became smaller, less expensive, more reliable, and thus more useful for other applications. As interest in Op Amps grew, their great flexibility became more apparent and they were more widely appreciated. The demands of the space age stimulated the development and growth of integrated circuits (ICs), which offer even greater reductions in cost and size and much-improved reliability. IC Op Amps are now available in a variety of small, inexpensive package types, and they offer the circuit designer virtually limitless possibilities for application. Practicing engineers and technicians must know and, in fact, be proficient with Op Amps and their applications even if their work environment only occasionally includes amplifications of small signals.

Workable circuits with Op Amps are easier to design and build than are most discrete equivalent circuits. However, the engineer or technician who intends to achieve reliable performance with Op Amps must acquire an understanding of the outstanding features and the shortcomings of Op Amps. The way a given Op Amp will behave in some known application can be predicted by its parameters. These parameters are provided by its manufacturer, and considerable emphasis on how to use such parameters is given in this book.

This book was developed to:

(1)   Provide an understandable and sufficiently detailed explanation of essential Op Amp theory and parameters for practicing engineers and technicians.

(2)   Serve as a text with plentiful examples and end-of-chapter problems.

(3)   Provide engineers, technicians, and students with a reference containing:

      (a)   A collection of Op Amp circuits with means of selecting their component values.

(b) Typical manufacturers' data sheets and tables of Op Amp types with their comparative characteristics.

Chapter 1 of this book has a review of junction and field-effect transistor fundamentals, principles of differential amplifiers, and the basic inner construction of the IC Op Amp. This first chapter can be omitted by those content with a "functional block" or "black box" approach to IC circuits.

Op Amps, and linear IC's in general, have a peculiar language of their own. Chapter 2 provides definitions and explanations of the essential Op Amp characteristics and compares their practical values to hypothetical ideal ones.

Chapters 3 through 7 provide detailed discussions of the significance of Op Amp parameters in practical circuits. Manufacturers' data sheets are provided and used frequently throughout.

Chapters 8 through 11 show and explain a variety of practical circuit applications. Methods of selecting their component values are especially emphasized.

Chapter 12 is about the current differencing (CD) amplifier, also known as the Norton amplifier or the Quad amplifier. While the CD amplifier is not exactly an Op Amp, its operation is similar enough to warrant discussion of it in this book. The CD amplifier offers space and cost savings in some applications that make it an interesting competitor of the Op Amp.

I want to thank the Op Amp manufacturers that generously supplied data sheets and application notes, my colleagues who offered constructive comments on the manuscript, my students who spent countless hours building and testing many of the Op Amp applications discussed here, and last but not least, my wife Angela and my daughters, who sacrificed recreational hours so that this book could be written.

GEORGE B. RUTKOWSKI

Handbook of
INTEGRATED-CIRCUIT
OPERATIONAL AMPLIFIERS

CHAPTER 1

DIFFERENTIAL AND OPERATIONAL AMPLIFIER CIRCUITS

The differential amplifier, as its name implies, amplifies the difference between two input signals. These input signals are applied to two separate input terminals, and their difference is called the differential input voltage V_{id}. The differential amplifier has two output terminals, and output signals can be taken from either output with respect to ground or across the two terminals themselves. The signals across the output terminals are usually highly amplified versions of the differential input voltages. Differential amplifiers, and variations of them, are found in many applications: electronic voltmeters, instrumentation, industrial controls, signal generators, computers, signal and dc amplifiers, to name only a few.

In this chapter we will first consider the construction and operating fundamentals of junction transistor and field effect transistor differential amplifiers. Then we will see how differential amplifiers can be cascaded, as they are in typical integrated circuits (ICs), to obtain very high gain. Finally, we will see how level-shifting circuitry is added to modify the two output terminals of a differential-type circuit to a single output from which a signal can be taken with respect to ground or a common point. Level shifting changes a differential amplifier into an operational amplifier (Op Amp).

1-1 Transistor Review

Since junction transistors and field effect transistors (FETs) are frequently used in discrete and IC differential and operational amplifiers, a review of these active devices is helpful if we expect to obtain even a general under-standing of the inner workings of linear ICs.

The junction transistor is a current-operated device with three terminals: an emitter E, a collector C, and a base B. In most applications, the relatively small base current I_B is controlled or varied by a signal source. The varying base current I_B in turn controls or varies a much larger collector current I_C and emitter current I_E.

Properly biased transistors are shown in Fig. 1-1. The term *bias* refers to the use of proper dc voltages on the transistor that are necessary to make it work. As shown in Fig. 1-1, the collector C is biased positively with respect to the emitter E on NPN transistors, whereas the collector C is normally negative with respect to the emitter E on PNP transistors. In either case, NPN or PNP circuit, the base-emitter junction of N- and P-type semi-conductors is forward biased. This means that V_{EE} is applied with a polarity that will admit current I_B across the base-emitter junction. The values of V_{EE}, V_s, R_E, and R_B dictate the value of I_B.

As base current flows, it causes a collector current flow, I_C, that is about β^* or h_{FE} times larger. That is,

$$I_C \cong \beta I_B \tag{1-1}$$

or

$$I_C \cong h_{FE} I_B \tag{1-2}$$

where β or h_{FE} is specified by the transistor manufacturer. In modern transis-tors, h_{FE} typically ranges from about 40 to 200. Thus the base current I_B is typically smaller than the collector current I_C by a factor between 40 and 200.

Note in Fig. 1-1 that in NPN transistor circuits, currents I_B and I_C flow into the transistor while I_E flows out. On the other hand, in PNP circuits, I_B and I_C flow out of the transistor while I_E flows in. By Kirchhoff's current law then

$$I_E = I_C + I_B \tag{1-3}$$

Since I_C is typically much larger than I_B, the base current I_B is often assumed negligible compared to either I_C or I_E. Therefore, Eq. (1-3) can be simplified to

$$I_E \cong I_C \tag{1-4}$$

---

* A transistor's beta (β) is about equal to its h_{FE}. In the symbol h_{FE}, the h means *hybrid* parameter, F means *forward* current transfer ratio, and E means that the *emitter* is common. The capital letters in subscript in h_{FE} specify that this is a dc parameter or dc beta.

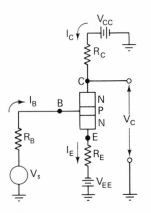

(a) The NPN transistor contains a P-type semiconductor between two N-type materials.

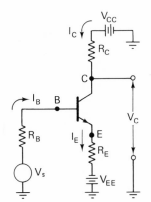

(b) Arrow on emitter points out on NPN transistor symbol.

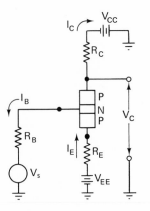

(c) The PNP transistor contains N-type semiconductor between two P-type materials.

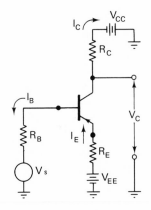

(d) Arrow on the emitter points in on PNP transistor symbol.

Fig. 1-1   Properly biased transistor circuits.

Variations of the input voltage V_s on any of the circuits of Fig. 1-1 will cause variations of I_B, which in turn varies I_C flowing through R_C. Thus the voltage across R_C and the output voltage V_C will vary too. Usually, the output variations of V_C are much larger than the input variations of V_s, which means that these circuits are capable of voltage gain—also called voltage amplification A_e.

The field effect transistor, FET, unlike the junction transistor, draws negligible current from the signal source V_s and therefore is referred to as a voltage-operated device. It has three terminals: a source S, a drain D, and a gate G. Properly biased FETs are shown in Fig. 1-2. Basically, variations of a voltage across the gate G and the source S cause drain current I_D and

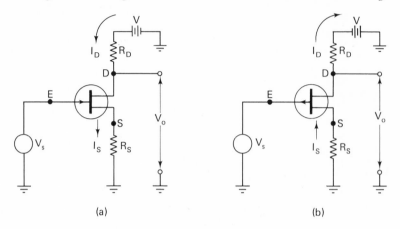

(a)   (b)

Fig. 1-2   Properly biased FETs: (a) N-channel FET and (b) P-channel FET.

source current I_S variations. Thus a varying input signal V_s causes a varying gate-to-source voltage V_{GS} which in turn varies I_D, the drop across R_D, and the output voltage V_o. The ratio of the change in drain current I_D to the change in gate-to-source voltage V_{GS} is the FET's transadmittance* y_{fs}, that is,

$$y_{fs} \cong \frac{\Delta I_D}{\Delta V_{GS}} \tag{1-5}$$

Because of the very high gate input resistance of the FET, negligible gate current flows, and therefore the drain and source currents are essentially equal, that is, $I_D = I_S$. FET amplifiers are capable of voltage gain, but not as much as are junction transistors. FETs are used where high input resistance is important, such as in some applications of differential and operational amplifiers. As we will see, some types of Op Amps on ICs use FETs in their first stage.

FETs are made in two general types: (1) Junction Field Effect Transistors, JFETs, whose input resistances are on the order of 10^6 to 10^9 ohms, and

---

* The forward transadmittance, sometimes called transconductance, is referred to by several symbols: $y_{fs}, g_{fs}, g_m, g_{21}$, etc.

whose symbols appear in Fig. 1-2, and (2) Metal-Oxide Silicon Field Effect Transistors, MOSFETs, which are also known as Insulated Gate Field Effect Transistors, IGFETs. MOSFETs or IGFETs have even higher gate input resistance values than do JFETs, typically 10^{10} to 10^{14} ohms. Some common symbols in use for MOSFETs are shown in Fig. 1-3.

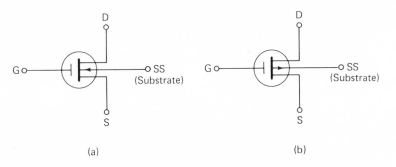

Fig. 1-3 Symbols used to represent MOSFETs or IGFETs: (a) N-channel, (b) P-channel.

1-2 The Differential Circuit

Figure 1-4a shows a simple junction transistor differential amplifier, and Fig. 1-4b, its typical symbol. If signal voltages are applied to input terminals 1 and 2, their difference V_{id} is amplified and appears as V_{od} across output terminals 3 and 4. Ideally, if both inputs are at the same potential with respect to ground or a common point, causing an input differential voltage $V_{id} = 0$ V, the differential output voltage $V_{od} = 0$ V too, regardless of the circuit's gain.

With input signal sources V_1 and V_2 applied as shown in Fig. 1-4, each base has a dc path to ground, and base currents I_{B_1} and I_{B_2} flow. If the average voltages and internal resistances of the signal sources V_1 and V_2 are equal, then equal base currents flow. This causes equal collector currents, $I_{C_1} = I_{C_2}$, and equal emitter currents, I_{E_1} and I_{E_2}, assuming that the transistors have identical characteristics. Since β or h_{FE} of a typical transistor is much larger than 1, the base current is very small compared to the collector or the emitter current, and therefore the collector and emitter currents are about equal to each other. See Eqs. (1-3) and (1-4). For example, in the circuit of Fig. 1-4, $I_{C_1} \cong I_{E_1}$ and likewise $I_{C_2} \cong I_{E_2}$. In this circuit, the current I_E in resistor R_E is equal to the sum of currents I_{E_1} and I_{E_2}. This current can

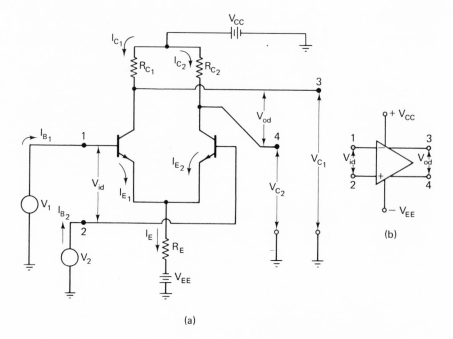

Fig. 1-4   (a) Simple differential amplifier; (b) symbol for the differential amplifier.

be approximated with the equation

$$I_E = \frac{V_{EE} - V_{BE}}{R_E + R_B/2h_{FE}} \tag{1-6}$$

where:

> V_{BE} is the dc drop across each forward-biased base-emitter junction, which is about 0.7 V in silicon transistors and about 0.3 V in germanium transistors, and
>
> R_B is the dc resistance seen looking to the left of either input terminal, 1 or 2. In the circuit of Fig. 1-4, R_B is the internal resistance of each signal source.

Since the dc source voltage V_{EE} is usually much larger than the base-emitter drop V_{BE} and since R_E is frequently much larger than $R_B/2h_{FE}$, Eq. (1-6) can be simplified to

$$I_E \cong \frac{V_{EE}}{R_E} \tag{1-7}$$

The significance of Eq. (1-7) is that the value of I_E is determined mainly by the values of V_{EE} and R_E and that if V_{EE} and R_E are fixed values, as they usually are, the current I_E is practically constant. Ideally, I_E should be very constant for reasons to be discussed later. The source V_{EE} and resistor R_E, or their equivalents, are often replaced with the symbol for a constant-current source as shown in Fig. 1-5.

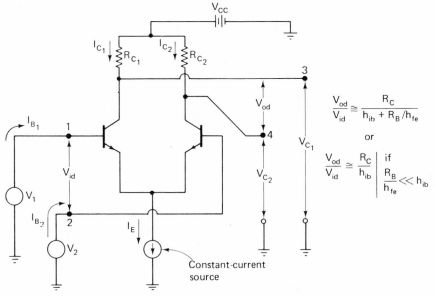

$$\frac{V_{od}}{V_{id}} \cong \frac{R_C}{h_{ib} + R_B/h_{fe}}$$

or

$$\frac{V_{od}}{V_{id}} \cong \frac{R_C}{h_{ib}} \quad \text{if} \quad \frac{R_B}{h_{fe}} \ll h_{ib}$$

Fig. 1-5 Constant-current source driving emitters on a differential amplifier.

The dc voltage to ground at each output terminal is simply the V_{CC} voltage minus the drop across the appropriate collector resistor. Thus the voltage at output 3 to ground is

$$V_{C_1} = V_{CC} - R_{C_1} I_{C_1} \tag{1-8a}$$

Similarly, the voltage at output 4 to ground is

$$V_{C_2} = V_{CC} - R_{C_2} I_{C_2} \tag{1-8b}$$

Their difference is the output differential voltage

$$V_{od} = V_{C_1} - V_{C_2} \tag{1-9}$$

just as the input differential voltage is

$$V_{id} = V_1 - V_2 \tag{1-10}$$

With I_E constant, as in the circuit of Fig. 1-5, the sum of the collector currents, $I_{C_1} + I_{C_2}$, and the sum of the emitter currents, $I_{E_1} + I_{E_2}$, are also constant. Thus if I_{C_1} is increased, I_{C_2} is forced to decrease, and vice versa. In other words, if the input voltage V_1 is made more positive, base current I_{B_1} increases, increasing I_{C_1} and the voltage drop across R_{C_1}. This in turn causes the voltage V_{C_1} at output 3 to ground to decrease. Furthermore, the increase in I_{C_1} forces I_{C_2} to decrease, which decreases the drop across R_{C_2} and increases the voltage V_{C_2} at output 4. We can therefore reason that, if V_1 becomes more positive than V_2, causing a larger differential input voltage V_{id}, then output 3 becomes more negative while output 4 becomes more positive, resulting in a larger differential output voltage V_{od}.

The ratio of the output voltage V_{od} to the input V_{id} is the differential voltage gain A_d of the differential amplifier. Therefore, combining Eqs. (1-9) and (1-10), we can show that

$$A_d = \frac{V_{od}}{V_{id}} = \frac{V_{C_1} - V_{C_2}}{V_1 - V_2} \qquad (1\text{-}11)$$

This differential gain can be estimated with the following equation:

$$A_d \cong \frac{R_C}{h_{ib} + R_B/h_{fe}} \qquad (1\text{-}12a)$$

where:

R_C is the resistance in series with each collector,
h_{ib}* is the dynamic emitter-to-base resistance of each transistor,
h_{fe} is the ac β, and
R_B is the resistance seen looking to the left of either input to ground (the internal resistance of either input signal source V_1 or V_2).

An approximate value of h_{ib} is sometimes provided on the manufacturer's data sheets, but it can also be estimated with the equation

$$\frac{25 \, \text{mV}}{I_C} \leq h_{ib} \leq \frac{50 \, \text{mV}}{I_C} \qquad (1\text{-}13)$$

If the internal resistance R_B of each signal source is small compared to the transistors' h_{ib} and if the h_{fe}'s are large, Eq. (1-12a) can be simplified to

$$A_d \cong \frac{R_C}{h_{ib}} \qquad (1\text{-}12b)$$

---

* The lower case letters in subscript in h_{ib} and h_{fe} specify that these are ac parameters and describe the dynamic characteristics of a transistor.

EXAMPLE 1-1

Referring to the circuit in Fig. 1-6 and assuming that both transistors are identical having a $\beta = 100$:

(a) Determine the approximate dc voltage to ground at each output and the output differential voltage when both inputs are 0 V (grounded).

(b) Find the approximate differential output voltage at some instant when input $V_1 = 2$ mV dc to ground and $V_2 = -1$ mV dc to ground. The internal resistances of the signal sources are negligible. Assume that for each transistor $V_{BE} \ll V_{EE}$.

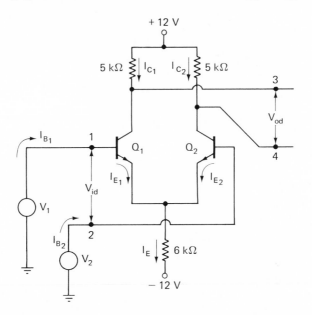

Fig. 1-6

Answer. (a) We can first estimate the current through R_E using Eq. (1-7):

$$I_E \cong \frac{12 \text{ V}}{6 \text{ k}\Omega} = 2 \text{ mA}$$

Since both inputs are at the same potential, both base currents are equal, causing equal collector currents. Because the sum of the collector currents must be about equal to I_E, each collector has about 1 mA in this case. The voltage V_{C_1} at output 3 is

$$V_{C_1} \cong 12\,\text{V} - 5\,\text{k}\Omega(1\,\text{mA}) = 7\,\text{V} \tag{1-8a}$$

Similarly, $V_{C_2} = 7\,\text{V}$ as determined with Eq. (1-8b). Therefore the output differential voltage is

$$V_{od} = 7\,\text{V} - 7\,\text{V} = 0\,\text{V} \tag{1-9}$$

(b) With different input voltages, a differential input exists. In this case

$$V_{id} = V_1 - V_2 = 2\,\text{mV} - (-1\,\text{mV}) = 3\,\text{mV} \tag{1-10}$$

This V_{id} is amplified by the circuit's differential gain A_d which can be estimated with the Eq. (1-12b). First we find the range of the dynamic emitter-to-base resistance, that is, the minimum

$$h_{ib} \cong \frac{25\,\text{mV}}{1\,\text{mA}} = 25\,\Omega$$

and the maximum

$$h_{ib} \cong \frac{50\,\text{mV}}{1\,\text{mA}} = 50\,\Omega$$

according to Eq. (1-13). Therefore, the maximum differential gain is

$$A_d \cong \frac{5000\,\Omega}{25\,\Omega} = 200$$

and the minimum gain is

$$A_d \cong \frac{5000\,\Omega}{50\,\Omega} = 100 \tag{1-12b}$$

Thus the differential output voltage can be as large as

$$V_{od} \cong 200(3\,\text{mV}) = 600\,\text{mV}$$

or as low as

$$V_{od} \cong 100(3\,\text{mV}) = 300\,\text{mV}$$

according to Eq. (1-11). The output differential voltage at terminal 3 is negative with respect to terminal 4, i.e., with input 1 more positive than input 2, collector current I_{C_1} is larger than I_{C_2}. This causes a larger drop across R_{C_1} than across R_{C2}, resulting in less voltage at terminal 3 than at terminal 4 with respect to ground.

A typical FET differential amplifier is shown in Fig. 1-7. It works in much the same way the transistor version does but has much larger resistance as seen looking into its input terminals 1 and 2. Integrated circuit differential and operational amplifiers are available with an FET input stage for applications where extremely large input resistance is needed or where significant dc input bias currents are undesirable. The differential voltage gain of the FET stage can be estimated with the equation

$$A_d = \frac{V_{od}}{V_{id}} \cong \frac{y_{fs}R_D}{R_Dy_{os} + 1} \qquad (1\text{-}14a)$$

where: y_{fs} is the forward transadmittance,
           y_{os} is the output admittance, and
           R_D is the resistance in series with the drain of either FET.

Frequently though, the product $R_Dy_{os} \ll 1$, and therefore the Eq. (1-14a) can be simplified to

$$A_d \cong y_{fs}R_D \qquad (1\text{-}14b)$$

An FET's parameters y_{fs} and y_{os} are provided on manufacturer's data sheets.

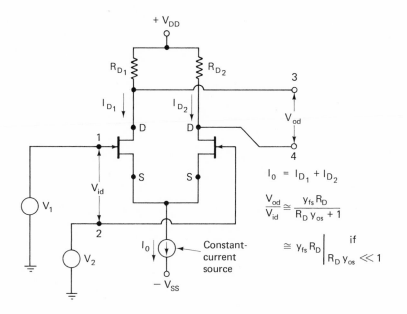

Fig. 1-7   FET differential amplifier.

1-3   Current-Source Biased Differential Amplifier

As mentioned previously, the emitters of a differential amplifier should ideally be driven by a constant-current source. A constant I_E gives the differential stage excellent *common-mode rejection* (*CMR*) capability. A differential amplifier has good *CMR* if it does not pass a *common-mode* (*CM*) voltage. A *CM* voltage is simply an input voltage that appears at both inputs simultaneously. For example, suppose that *both* inputs, 1 and 2, in the circuit of Fig. 1-5, are increased from 0 V to 10 mV above ground. The 10 mV, then, is a common-mode voltage. In this case, it makes both bases more positive, and both transistors attempt to turn on harder. That is, the collector currents of both transistors will attempt to increase, but since their sum is about equal to I_E (a constant), neither of the collector currents increase and the voltages to ground at outputs 3 and 4 remain essentially constant. Thus if a common-mode voltage is applied, it is not amplified; only differential voltages V_{id} are. In practice, no differential stage has ideal (infinite)

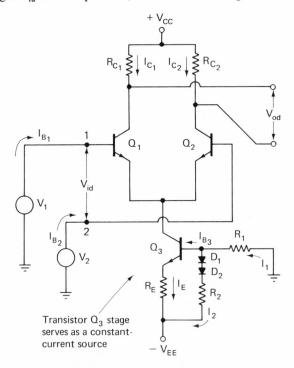

Fig. 1-8 Current-source biased differential amplifier.

CMR, and some CM voltage appears across the output terminals if a CM voltage is applied to its inputs. Typically though, the output common-mode voltage V_{cmo} is much smaller than the input common-mode voltage V_{cmi}. One reason the practical CMR is not ideal is that the current I_E driving the emitters is not perfectly constant. Generally, the more constant I_E is, the better the CMR.

In practice, the constancy of I_E is improved by driving the emitters with a transistor stage such as in the circuit of Fig. 1-8. The current through transistor Q_3 is very constant even if a varying V_{cmi} is applied or if a change in temperature occurs. The base of Q_3 is biased with the voltage divider containing components R_1, D_1, D_2, and R_2. The diodes D_1 and D_2 help to hold I_E constant even though the temperature changes. Note that current I_1 flows to the node at the base of Q_3 and then divides into paths I_2 and I_{B_3}. If the temperature of Q_3 increases, its base-emitter voltage V_{BE} decreases about 2.5 mV/°C. This reduced V_{BE} tends to raise the voltage drop across R_E and the current I_E. However, the voltage drops across D_1 and D_2 likewise decrease, causing a greater portion of I_1 to contribute to I_2, that is, to flow down through D_1 and D_2. This causes I_{B_3} to decrease, which prevents any significant increase in I_E.

1-4   Multistage Differential Circuits

Cascading amplifier stages provides an overall (total) gain that is the product of the individual stage gains. In many applications, such as in linear ICs, differential amplifiers are cascaded for large total gain as shown in Fig. 1-9. A point to note is that the outputs of the first stage are *directly coupled* to the inputs of the second stage. This presents a problem that requires different biasing methods in these individual stages. For reasons that are discussed in detail in a later chapter, the input (first) stage must be able to take input voltages V_1 and V_2 that might vary positively or negatively with respect to ground, within limits, and have as a result of these input variations, continuous variations of its differential output voltage v_{od}. This means that the transistors or FETs of the input stage should not saturate or cut off with inputs in the neighborhood of about 0 V to ground or common. A transistor is saturated when the current into its base is so large as to reduce its V_{CE} voltage to about zero. It is cut off when its base current is essentially zero causing the transistor to behave as an open between its collector and emitter. The constant-current source in the input stage, whether it is a

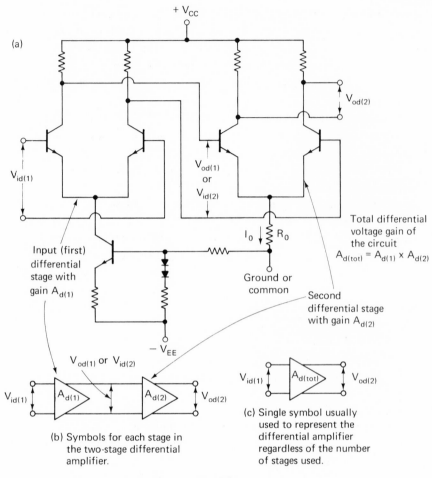

(a)

$+ V_{CC}$

$V_{od(2)}$

$V_{id(1)}$

$V_{od(1)}$
or
$V_{id(2)}$

I_0 R_0

Total differential
voltage gain of
the circuit
$A_{d(tot)} = A_{d(1)} \times A_{d(2)}$

Input (first)
differential
stage with
gain $A_{d(1)}$

Ground or
common

Second
differential stage
with gain $A_{d(2)}$

$- V_{EE}$

$V_{od(1)}$ or $V_{id(2)}$

$V_{id(1)}$ $A_{d(1)}$ $A_{d(2)}$ $V_{od(2)}$

(b) Symbols for each stage in
the two-stage differential
amplifier.

$V_{id(1)}$ $A_{d(tot)}$ $V_{od(2)}$

(c) Single symbol usually
used to represent the
differential amplifier
regardless of the number
of stages used.

Fig. 1-9

transistor stage or a resistor R_E and dc source V_{EE}, holds the emitters in
Fig. 1-5 and the sources S in Fig. 1-7 at about ground potential. This allows
us to use input voltages that can be varied somewhat around ground potential
because the bases must normally be at approximately the same voltage to
ground as are the emitters.

The second stage, however, has a considerable dc component applied to
each of its inputs (bases) since the collectors of the first stage are above
ground (have a dc voltage to ground) and are applied to the inputs of the
second stage. To avoid saturation of the transistors of the second stage, their
emitters must be above ground by about the same potential as their bases.

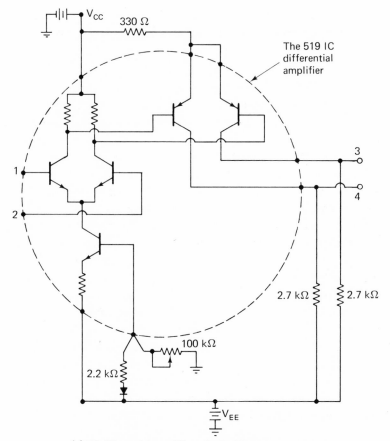

V_{CC}

330 Ω

The 519 IC
differential
amplifier

3

1

2

4

2.7 kΩ   2.7 kΩ

100 kΩ

2.2 kΩ

V_{EE}

(a) IC differential amplifier wired with external components
to have about 0 V to ground at each output terminal 3 and 4.

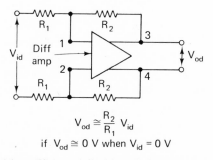

R_1 R_2

3

V_{id} Diff   1
amp

V_{od}

2

4

R_1 R_2

$$V_{od} \cong \frac{R_2}{R_1} \, V_{id}$$

if $V_{od} \cong 0$ V when $V_{id} = 0$ V

(b) IC differential amplifier symbol with typical externally connected
input resistors R_1 and feedback resistors R_2

Fig. 1-10

This is accomplished by the use of resistor R_0, one end of which is grounded instead of terminated to a negative dc bias source. The differential gain of this two-stage combination is

$$A_{d(\text{tot})} = \frac{V_{od(2)}}{V_{id(1)}} = A_{d(1)} \times A_{d(2)}$$

Of course, with the emitters of both transistors in the second stage above ground, the collectors are even more positive to ground considering each transistor's collector-to-emitter drop.

The dc voltages to ground, at the outputs of the two-stage differential amplifier in Fig. 1-9, are usually inconvenient and undesirable for reasons discussed in later chapters. Therefore, two differential stages are often wired so that the output terminals of the second stage are very nearly 0 V with respect to ground when both inputs of the first stage are at 0 V (grounded). Such a connection is shown in Fig. 1-10a. The components in the circle are typical of those contained in an IC differential amplifier. With each input 1 and 2 grounded, the externally connected 100 kΩ potentiometer is adjusted so that the drop across each external 2.7 kΩ resistor is equal in value to the V_{EE} voltage which puts both outputs 3 and 4 at about 0 V with respect to ground. Of course, these outputs swing above or below ground when differential input voltages V_{id} are applied to inputs 1 and 2. With the outputs at about ground potential, additional externally connected resistors are easily and often used to stabilize the differential gain. The input resistors R_1 and the feedback resistors R_2 shown in Fig. 1-10b are examples of this. Feedback is thoroughly discussed in another chapter.

1-5   Level Shifting with Intermediate Stage

In the previous section we learned that it is often convenient to have both outputs of a differential amplifier near or at ground potential. Also convenient, for reasons discussed later, is an amplifier with a differential input (two input leads) but with a single-ended output (one output lead). The output signals can then be taken off this single terminal with respect to ground or a common point. Such a configuration—differential input and single-ended output—is basically an operational amplifier (Op Amp). Op Amps and their applications are thoroughly discussed in this book and are viewed mainly from an external point of view. In this chapter, though, we can look at some internal details and see how an Op Amp is built by adding intermediate and output circuitry to the two-stage differential amplifier.

Since both outputs of the second stage in Fig. 1-9 are normally well above ground, neither of them will serve as an Op Amp output terminal. However, either output can be shifted down to about 0 V to ground with appropriate voltage-divider circuitry. For example, an output of the second differential stage can work into an emitter-follower stage as shown in Fig. 1-11a. Thus a positive voltage at point B can cause 0 V to ground at point E with proper selection of components. That is, if the sum of the voltage drops across the transistor Q and resistor R_a equals the V_{CC} voltage, then the resistor R_b must drop the V_{EE} voltage, and point E is at 0 V with respect to ground. Better results are obtained by using a transistor stage Q_2 as a constant-current source, as shown in Fig. 1-11b, instead of the combination of R_b and V_{EE}.

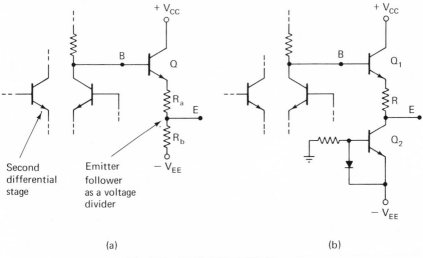

(a)   (b)

Fig. 1-11   Level-shifting techniques.

In this case also, point E is near ground potential though point B is well above ground. If point B is driven more positively by the preceding differential stage, transistor Q_1 will drop less voltage and Q_2 will drop more, causing point E to swing positively. Then if point B swings negatively, Q_1's drop increases and Q_2's drop decreases, causing point E to swing negatively.

1-6   The Output Stage and the Complete Operational Amplifier

Though the output terminal E of the level-shifting stage is not as positive as are the collectors of the preceding differential stage, this output E is not

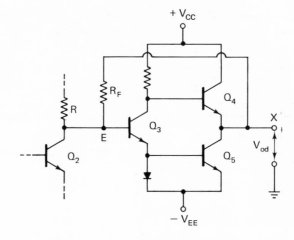

Fig. 1-12 Output stage of a typical Op Amp.

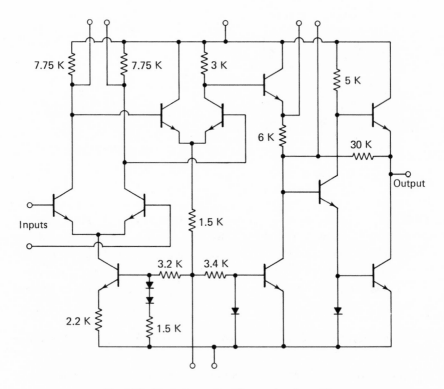

Fig. 1-13 Typical complete IC operational amplifier.

usually used as the output terminal of the Op Amp. An additional output stage, such as in Fig. 1-12, is used. The transistor Q_3 amplifies the signal at point E and drives the *totem-pole* arrangement of transistors Q_4 and Q_5. This output stage not only provides zero or near zero voltage to ground at its output X when the differential input V_{id} to the first stage is zero, but also provides large peak-to-peak output signal capability when an input V_{id} is applied. In other words, the output X can swing positively to about the V_{CC} voltage and negatively to about the V_{EE} voltage. The resistor R_F provides feedback which stabilizes the output stage, preventing the voltage at X from drifting with temperature changes.

A typical complete operational amplifier is shown in Fig. 1-13. By examining this circuit we can see the stages discussed previously. While specific inner design of IC Op Amps is continually evolving, including FET or Darlington pair inputs, internal compensations, input voltage and output current limiter, and many others, the circuits of this chapter give us a basic understanding of the inner operation of the IC Op Amp. Other components and circuitry contained in some types of Op Amps will be introduced as their relevance becomes apparent.

REVIEW QUESTIONS

1. In which lead of a transistor is the current much smaller than in the other two leads?

2. In which leads of a transistor are the currents approximately equal?

3. The same current flows in which leads of an FET?

4. The drain current in an FET circuit is controlled by a voltage across what leads?

5. What does a differential amplifier do?

6. The junction transistor differential amplifier has (*higher*) (*lower*) input resistance and a (*higher*) (*lower*) voltage gain than does the FET version.

7. What is the purpose of cascading differential amplifiers?

8. From the external point of view, how does an Op Amp differ from a differential amplifier?

9. What is the purpose of an intermediate stage following the second differential amplifier as shown in Fig. 1-13?

10. Generally, over what approximate range is the output voltage of the Op Amp in Fig. 1-13 able to swing?

PROBLEMS

1. If a transistor's collector current $I_C = 0.5$ mA and its $h_{FE} = 80$, what is its approximate base current I_B?

2. If in the circuit of Fig. 1-4 both inputs are grounded, $V_{EE} = V_{CC} = 10$ V, $R_{C_1} = R_{C_2} = 7.2$ kΩ, and $R_E = 10$ kΩ, what is the voltage at each collector with respect to ground, and what is the differential output voltage? (Assume that the transistors are identical.)

3. Referring to the circuit described in the previous problem, what approximate voltage can we expect at each output to ground, and what is the differential output voltage if +400 mV dc is applied to each input?

4. Referring to the circuit described in Problem 2, approximately what dc base current flows in each input terminal if the β of each transistor is 200?

5. What is the approximate range of voltage gain of the circuit described in Problem 2 assuming that the internal resistances of V_1 and V_2 are negligible?

6. If in the circuit of Fig. 1-7, $I_o = 0.4$ mA, $V_{DD} = 12$ V, and $R_{D_1} = R_{D_2} = 20$ kΩ, what is the approximate drain-to-ground voltage on each FET when both gates are grounded? (Assume that the FETs are identical.)

7. Suppose that we are to adjust both outputs of the circuit in Fig. 1-10a to 0 V with respect to ground while both inputs are grounded. By adjusting the 100-kΩ potentiometer, we obtain 0 V at output 3, but then we have +1.5 V at output 4. With further adjustment, output 4 becomes 0 V, but then output 3 acquires −1.5 V to ground. What change in external wiring or components would you recommend?

CHAPTER 2

CHARACTERISTICS OF OPERATIONAL AMPLIFIERS AND THEIR POWER SUPPLY REQUIREMENTS

The term *operational amplifier* refers to a high-gain dc amplifier that has a differential input (two input leads) and a single-ended output (one output lead). The signal output voltage V_o is larger than the differential input signal across the two inputs by the gain factor of the amplifier. See Fig. 2-1. Op Amps have characteristics such as high input resistance, low output resistance, high gain, etc., that make them highly suitable for many applications, a number of which are shown and discussed in later chapters. By examining some applications and comparing the characteristics of typical Op Amps, we will see that some types are apparently better than others. The overall most desirable characteristics that Op Amp manufacturers strive to obtain in their products are *ideal characteristics*. While some of these ideal characteristics are impossible to obtain, we often assume that they exist to develop circuits and equations that work perfectly on paper. These circuits and equations then work very well with practical Op Amps, provided their characteristics are not too far from the ideal values. Actual Op Amp characteristics are more or less ideal, relative to conditions external to the Op Amp: signal source resistance, load resistance, amount of feedback used, etc. Some of the more important characteristics, ideal and practical values, are given and defined in the following sections.

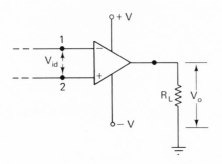

Fig. 2-1 Typical Op Amp symbol. $+V$ is the positive dc power supply voltage; $-V$ is the negative dc power supply voltage; $A_{VOL} = V_o/V_{id}$.

2-1 Open-Loop Voltage Gain* A_{VOL}

The open-loop voltage gain A_{VOL} of an Op Amp is its *differential* gain under conditions where no negative feedback is used as shown in Figs. 2-1 and 2-2. Ideally its value is infinite, that is, the mythical ideal Op Amp has an open-loop voltage gain.

$$A_{VOL} = \frac{V_o}{V_{id}} = -\infty \qquad (2\text{-}1a)$$

or

$$A_{VOL} = \frac{V_o}{V_1 - V_2} = -\infty \qquad (2\text{-}1b)$$

The negative sign means that the output V_o and the input V_{id} are out of phase. The concept of an infinite gain is difficult to visualize and impossible to obtain. The important point to understand is that the Op Amp's output voltage V_o should be very much larger than its differential input V_{id}. To put

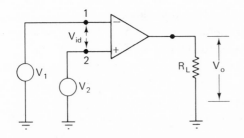

Fig. 2-2 Op Amp with input voltages V_1 and V_2 whose difference is V_{id}. Power supply voltages $+V$ and $-V$ are assumed if not shown.

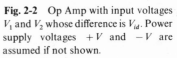

---

* The open-loop voltage gain is referred to with a variety of symbols: A_{VOL}, A_d, A_{EOL}, A_{VD}, etc.

it another way, the input V_{id} should be infinitesimal compared to any practical value of output V_o. Open-loop gains A_{VOL} range from about 5000 (about 74 dB) to 100,000 (about 100 dB), that is,

$$5000 \leq A_{VOL} \leq 100{,}000 \qquad (2\text{-}2a)$$

or

$$74 \text{ dB} \leq A_{VOL} \leq 100 \text{ dB} \qquad (2\text{-}2b)$$

with popular types of Op Amps. The fact that the output V_o is 5000 to 100,000 times larger than the differential input V_{id} does not mean that V_o can actually be very large. In fact, its positive and negative peaks are limited to values a little less than the positive and negative supply voltages being used to power the Op Amp. Since the dc supply voltages for IC Op Amps are usually less than 20 V, the peaks of the output voltage V_o are less than 20 V. This fact coupled with the high gain factor of the typical Op Amp makes the voltage V_{id} across the inputs 1 and 2 very small. Of course, the larger the open-loop gain A_{VOL}, the smaller V_{id} is in comparison to any practical value of V_o. Thus, since

$$A_{VOL} = \frac{V_o}{V_{id}} \qquad (2\text{-}3a)$$

or

$$A_{VOL} = \frac{V_o}{V_1 - V_2} \qquad (2\text{-}3b)$$

then

$$V_{id} = \frac{V_o}{A_{VOL}}$$

and therefore*

$$\lim_{A_{VOL} \to \infty} V_{id} = 0$$

Thus if the gain A_{VOL} in the expression V_o/A_{VOL} is very large, the value of this expression, which is V_{id}, must be relatively very small. In fact, V_{id} is usually so small that we can assume there is practically no potential difference between the inverting and noninverting inputs.

---

* As the value of open-loop gain A_{VOL} approaches infinity, the value of differential input voltage V_{id} approaches zero if $V_o \neq 0$.

2-2   Output Offset Voltage V_{oo}

The output offset voltage V_{oo} of an Op Amp is its output voltage to ground or common under conditions when its differential input voltage $V_{id} = 0$ V. Ideally $V_{oo} = 0$ V. In practice, due to imbalances and inequalities in the differential amplifiers within the Op Amp itself (see Fig. 1-13), some output offset voltage V_{oo} will usually occur even though the input $V_{id} = 0$ V. In fact, if the open-loop gain A_{VOL} of the Op Amp is high and if no feedback is used, the output offset can be so large as to saturate or nearly saturate the output. In such cases, the output voltage is either a little less than the positive source voltage $+V$ or the negative source $-V$. This is not as serious as it first appears because corrective measures are fairly simple to apply and because the Op Amp is seldom used without some feedback. When corrective action is taken to bring the output to 0 V when the applied differential input signal is 0 V, the Op Amp is said to be *balanced* or *nulled*.

If both inputs are at the same finite potential, causing $V_1 - V_2 = V_{id} = 0$ V, and if the output $V_o = 0$ V as a result, the Op Amp is said to have an ideal *common-mode rejection* (*CMR*). Though most practical Op Amps have a good *CMR* capability, they do pass some *common-mode* (*CM*) voltage to the load R_L. Typically though, the load's common-mode voltage V_{cmo} is hundreds or even thousands of times *smaller* than the input common-mode voltage V_{cmi}. A thorough discussion of the practical Op Amp's *CMR* capability and its applications is given in a later chapter.

2-3   Input Resistance R_i

The Op Amp's input resistance R_i is the resistance seen looking into its inputs 1 and 2 as shown in Fig. 2-3. Ideally $R_i = \infty$ Ω. Practical Op Amps' input resistances are not infinite but instead range from less than 5 kΩ to over 20 MΩ, depending on type. Though resistances in the low end of this range seem a bit small compared to the desired ideal (∞ Ω), they can be quite large compared to the low internal resistances of some signal sources

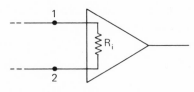

Fig. 2-3 Input resistance R_i is the resistance seen looking into the inputs 1 and 2.

commonly used to drive the inputs of Op Amps. Generally, if a high-resistance signal source is to drive an Op Amp, the Op Amp's input resistance should be relatively large. As we will see later, the *effective* input resistance R_i' is made considerably larger than the manufacturer's specified R_i by wiring the Op Amp to have negative feedback. Manufacturers usually specify their Op Amps' input resistances as measured under open-loop conditions (no feedback). In most linear applications, Op Amps are wired with some feedback and this can improve (increase) the effective resistance seen by the signal source driving the Op Amp.

2-4   Output Resistance R_o

With a differential input signal V_{id} applied, the Op Amp behaves like a signal generator as the load connected to the output sees it. As shown in Fig. 2-4, the Op Amp can be shown as a signal source generating an open-circuit voltage of $A_{VOL}V_{id}$ and having an internal resistance of R_o. This R_o is the Op Amp's output resistance and ideally should be $0\,\Omega$. Obviously, if $R_o = 0\,\Omega$ in the circuit of Fig. 2-4, all of the generated output signal $A_{VOL}V_{id}$

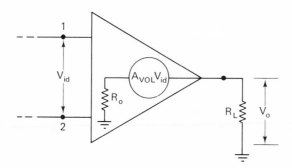

Fig. 2-4   Output resistance R_o tends to reduce output voltage V_o.

appears at the output and across the load R_L. Depending on the type of Op Amp, specified output resistance R_o values range from a few ohms to a few hundred ohms and are usually measured under open-loop (no feedback) conditions. Fortunately, the *effective* output resistance R_o' is reduced considerably when the Op Amp is used with some feedback. In fact, in most applications using feedback, the effective output resistance of the Op Amp is very nearly ideal $(0\,\Omega)$. The effect feedback has on output resistance is discussed in more specific terms later.

2-5   Bandwidth *BW*

The *bandwidth BW* of an amplifier is defined as the range of frequencies at which the output voltage does not drop more than 0.707 of its maximum value while the input voltage *amplitude* is constant. Ideally, an Op Amp's bandwidth $BW = \infty$. An infinite bandwidth is one that starts with dc and extends to infinite cycles/second (Hz). It is indeed idealistic to expect such a bandwidth from any kind of amplifier. Practical Op Amps fall far short of this ideal. In fact, limited high-frequency response is a shortcoming of Op Amps. While some types of Op Amps can be used to amplify signals up to a few megahertz, they require carefully chosen externally wired compensating components. Most general purpose types of IC Op Amps are limited to less than 1 MHz bandwidth and more often are used with signals well under a few kilohertz, especially if any significant gain is expected of them. The limited high-frequency response of Op Amps is not a serious matter with most types of instrumentation in which Op Amps are used extensively.

2-6   Response Time

The response time of an amplifier is the time it takes the output of an amplifier to change after the input voltage changes. Ideally, response time = 0 seconds, that is, the output voltage should respond instantly to any change on the input. Figure 2-5 shows an Op Amp's typical output-voltage response to a step input voltage when wired for unity gain. Manufacturers specify a *slew rate* that gives the circuit designer a good idea of how quickly a given Op Amp responds to changes of input voltage. Note that the time scale in

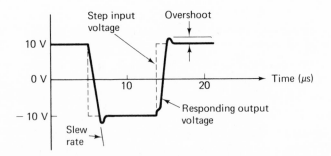

Fig. 2-5   Typical Op Amp response to a step input voltage when wired as a voltage follower (unity gain).

Fig. 2-5 is in microseconds (μs), which indicates that the typical Op Amp responds quickly though not instantly. Also note that the output voltage, when changing, overshoots the level it eventually settles at. Overshoot is the ratio of the amount of overshoot to the steady-state deviation expressed as a percentage. For example, if we observe the responding output voltage of Fig. 2-5 on an oscilloscope whose vertical deflection sensitivity is set at 10 V/cm, and the amount of overshoot measures 0.2 cm, the percentage of overshoot is

$$\frac{\text{amount of overshoot}}{\text{steady-state deflection}} \times 100 = \frac{0.2\,\text{cm}}{2\,\text{cm}} \times 100 = 10\%$$

From time to time we will refer to the various *ideal* characteristics, and therefore the following summary of them will be helpful:

(1) Open-loop voltage gain $A_{VOL} = \infty$.
(2) Output offset voltage $V_{oo} = 0$ V.
(3) Input resistance $R_i = \infty\ \Omega$.
(4) Output resistance $R_o = 0\ \Omega$.
(5) Bandwidth $BW = \infty$ Hertz.
(6) Response time $= 0$ seconds.

2-7   Power Supply Requirements

In many applications, the Op Amp's output voltage V_o must be capable of swinging in both positive and negative directions. In such applications, the Op Amp requires two source voltages: one positive $(+V)$ and the other negative $(-V)$ with respect to ground or a common point. These dc source voltages must be well filtered and regulated, otherwise the Op Amp's output voltage will vary with the power supply variations. The output voltage of an Op Amp varies more or less with power supply variations, depending on its closed-loop* voltage gain and sensitivity factor S which is usually specified on the manufacturer's data sheets. An ideal Op Amp has a sensitivity factor $S \doteq 0$, which means that power supply voltage variations have no effect on its output. Practical Op Amps, however, are affected by changes in the supply voltages, and therefore regulated supplies are used to keep Op Amp outputs responsive to differential input voltages only.

---

* The closed-loop gain of an Op Amp is its gain when a feedback loop is used.

Some power supplies use Op Amps as part of their voltage regulating system, and these are discussed in a later chapter. If the current drain is not too large and if it is relatively constant, simple zener-diode-regulated* power supplies such as in Fig. 2-6 can be used. Some types of IC Op Amps can work with dc source voltages up to and over ± 20 V, but usually values under ± 15 V are recommended.† Thus, in practice, the output voltages of the power supplies in Fig. 2-6 are ± 15 V or under. The circuits in Fig. 2-6c

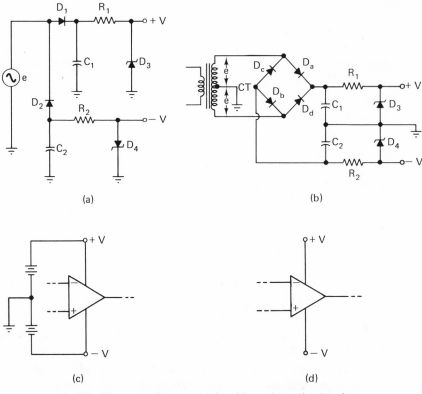

(a)   (b)

(c)   (d)

Fig. 2-6 Power supplies capable of positive and negative dc voltages to ground or a common point: (a) half-wave zener-regulated circuit; (b) full-wave zener-regulated circuit; (c) battery symbols sometimes used to represent plus and minus source voltages; (d) common method of showing dc source voltages.

---

*Zener diodes are also called *regulator* diodes on typical data sheets.

†Some hybrid Op Amps, containing discrete components and ICs, are designed to work with dc supply voltages over 100 V and hundreds of volts peak-to-peak output capability.

and d show how the dc power sources are sometimes drawn. Usually these dc sources and the pins to which they are connected are not shown but are assumed to be there. See Figs. 2-2, 2-3, and 2-4.

Generally, the ac voltages e in the circuits of Fig. 2-6a and b are selected so that their peaks exceed the actual required dc outputs $+V$ or $-V$ by 10% to 40%, depending on how unregulated this ac voltage e is. Capacitors C_1 and C_2 charge to approximately the peak of ac input e because of the rectifying action of diodes D_1 and D_2 or of diodes D_a through D_d. If voltage e is sinusoidal, each capacitor's voltage is about $e \sqrt{2}$ volts, where e is the rms value. The difference between each capacitor's voltage and its respective zener voltage appears across the resistor. Thus the difference between the voltages across C_1 and D_3 appears across R_1. Similarly, the drop across R_2 is the difference between the voltages across C_2 and D_4. The zeners can normally conduct a varying current and hold a fairly constant voltage drop. This characteristic enables them to act as voltage regulators. Thus if the ac input voltage e increases or decreases, as an unregulated ac source typically does, each zener draws more or less current, respectively. This increases or decreases the drop across each resistor, but the zener and output voltages remain quite constant. On the other hand, if voltage e is constant but the loads connected to the $+V$ and $-V$ output terminals draw more or less current, the zeners respond by drawing less or more current, respectively, to keep the drops across the resistors and zeners quite constant. The point is, the zeners have reasonably constant voltage drops across them even though the ac input voltage e and/or the load resistances vary.

The component values for the regulated power supply circuits in Fig. 2-6 can be selected as follows:

(1) Select the source of ac voltage e (usually a transformer secondary) so that its peak is about 10% to 40% larger than the actual $+V$ and $-V$ outputs required. The dc voltage across each filter capacitor is about equal to the peak of the ac voltage e.

(2) Select the series dropping resistors with the equation

$$R < \frac{E_{min} - V_z}{I_{L(max)}} \qquad (2-4)$$

where:

E_{min} is the minimum dc across each of the capacitors C_1 and C_2, V_z is the zener voltage of D_3 or D_4 (assuming that the circuit is symmetrical and that both zeners are the same type),

$I_{L(max)}$ is the maximum current drawn by each of the loads on outputs $+V$ and $-V$, and

R is the approximate value of each resistor R_1 and R_2.

After the value of R is calculated, the actual values of R_1 and R_2 are selected to be slightly less. This allows for some current flow in each of the zeners at all times and prevents loss of regulation.

(3) The power rating of each series dropping resistor must exceed

$$P_{R(max)} = \frac{(E_{max} - V_z)^2}{R} \tag{2-5}$$

where E_{max} is the maximum dc voltage across each capacitor C_1 and C_2.

(4) The power rating of each zener must exceed

$$P_{z(max)} = \left[\frac{E_{max} - V_z}{R} - I_{L(min)} \right] V_z \tag{2-6}$$

where: R is the actual value of each of the resistors R_1 and R_2 used, and
$I_{L(min)}$ is the minimum current each load draws from the dc output terminals $+V$ and $-V$.

(5) As a rule of thumb, the maximum secondary rms current rating of the transformer, or the output-current capability of any source of the ac voltage e, should exceed the maximum dc current flow through R_1 and R_2. More specifically, the current capability of the source e must exceed four times the maximum current in R_1 or R_2 in the circuit of Fig. 2-6a. In Fig. 2-6b, the transformer secondary-current (I_s) capability should exceed 1.8 times the maximum current in R_1 or R_2. Thus for the circuit in Fig. 2-6a,

$$I_s > 4I_{R(max)} = 4 \left[\frac{E_{max} - V_z}{R} \right] \tag{2-7a}$$

And for the circuit in Fig. 2-6b,

$$I_s > 1.8I_{R(max)} = 1.8 \left[\frac{E_{max} - V_z}{R} \right] \tag{2-7b}$$

(6) The PIV (peak inverse voltage) ratings of the rectifier diodes, D_1 and D_2 or D_a through D_d, must exceed twice the maximum peak of voltage e. Thus if e is the rms value, then

$$PIV > 2(e\sqrt{2}) \tag{2-8}$$

and generally, their current ratings should exceed the maximum expected secondary rms current as determined with the appropriate version of Eq. (2-7).

(7) At 60 Hz, for reasonable filtering capability, each of the filter capacitors should have a capacitance of at least

$$C \cong 50 \text{ ms} \left(\frac{I_{R(\max)}}{V_z} \right) \tag{2-9a}$$

for the half-wave circuit such as in Fig. 2-6a, or at least

$$C \cong 25 \text{ ms} \left(\frac{I_{R(\max)}}{V_z} \right) \tag{2-9b}$$

for the full-wave circuit such as in Fig. 2-6b.

EXAMPLE 2-1

Suppose a transformer is the source of the ac voltage e in the circuit of Fig. 2-6a, and that its secondary is specified to be 12 V rms when the primary is 115 V rms. If the output voltages are to be ± 12 V dc, each of the loads on the $+V$ and $-V$ outputs draws 0 to 40 mA, and the primary voltage varies from 110 V to 125 V rms. Determine:
  (a) the resistances of R_1 and R_2 and their power ratings,
  (b) the zeners' voltages V_z and the required power rating of each,
  (c) the transformer's approximate required secondary-current capability,
  (d) the rectifier diodes' PIV and current ratings, and
  (e) the filter capacitors' approximate values.

Answer. This transformer's turns ratio $N_p/N_s = 115 \text{ V}/12 \text{ V} \cong 9.6$. Thus when the primary voltage is 110 V (minimum), the minimum secondary voltage $e_{\min} \cong 110 \text{ V}/9.6 \cong 11.5$ V rms, and therefore the minimum dc voltage across each capacitor is

$$E_{\min} \cong 11.5\sqrt{2} \cong 16.2 \text{ V}$$

When the primary voltage is 125 V, the maximum secondary voltage $e_{\max} \cong 125 \text{ V}/9.6 \cong 13$ V rms, which causes a maximum dc voltage across each capacitor of

$$E_{\max} = 13\sqrt{2} \cong 18.4 \text{ V}$$

Since we require ± 12-V outputs, each zener is selected so that $V_z \cong 12$ V; therefore

(a)
$$R < \frac{16.2 - 12}{40 \text{ mA}} = \frac{4.2 \text{ V}}{40 \text{ mA}} \cong 105 \, \Omega \tag{2-4}$$

Since R_1 and R_2 are to be slightly less than this calculated R, we look for available resistors under $105 \, \Omega$. Standard $100 \, \Omega$ resistors will probably be unsuitable, considering their normal tolerances. For example, a resistor rated at $100 \, \Omega$ with 20% or 10% tolerance can have a resistance of more than $105 \, \Omega$. A commonly available, next-size-smaller, standard value is $75 \, \Omega$; therefore let $R_1 = R_2 = 75 \, \Omega$ in this case. This leads to each resistor's maximum power dissipation:

$$P_{R(\max)} = \frac{(18.4 - 12)^2}{75 \, \Omega} = \frac{6.4^2}{75 \, \Omega} \cong 0.55 \text{ W} \tag{2-5}$$

A $\frac{3}{4}$-watt or 1-watt rating will probably work well in most environments.

    (b) We've already selected each zener's voltage to be 12 V. Each one's maximum power dissipation is

$$P_{z(\max)} \cong \left[\frac{18.4 - 12}{75} - 0 \right] 12 \cong 1 \text{ W} \tag{2-6}$$

Zeners with a power rating of 1 watt or more should be used, depending on the ambient temperature. Maximum power dissipation versus temperature curves, or derating factors, are provided by the zener manufacturers.

    (c) With the values of $E_{\max}$, V_z, and R known, the maximum current in R may be calculated:

$$I_{R(\max)} = \frac{18.4 - 12}{75} = \frac{6.4 \text{ V}}{75 \, \Omega} \cong 85.4 \text{ mA}$$

Therefore, by Eq. (2-7a), the required current capability of the secondary is

$$I_s \cong 4(85.4 \text{ mA}) \cong 342 \text{ mA}$$

Like the transformer's secondary, each rectifier diode should be capable of at least 342 mA in this case.

    (d) $PIV > 2(13 \text{ V} \sqrt{2}) = 36.8 \text{ V}$ (2-8)

    (e) Since this circuit is a half-wave type, a filter capacitance

$$C = 50 \text{ ms} \left[\frac{I_{R(\max)}}{V_z} \right] = 50 \text{ ms} \left[\frac{85.4 \text{ mA}}{12 \text{ V}} \right] \cong 350 \, \mu\text{F}$$

will keep the ac ripple output negligible for most practical purposes.

REVIEW QUESTIONS

Fill in the blanks of the next ten problems with one of the six following terms:

    (I)   infinite   (IV) relatively small
    (II) zero   (V) decrease
    (III) relatively large   (VI) increase

1. The ideal Op Amp's output resistance is _____ ohms.

2. The ideal Op Amp's open-loop voltage gain is _____.

3. The ideal input resistance of an Op Amp is _____ ohms.

4. The ideal Op Amp's bandwidth is _____ hertz.

5. The ideal response time of an Op Amp is _____ seconds.

6. Compared to the internal resistance of the signal source, the practical Op Amp's effective input resistance should be _____.

7. Compared to the load resistance, the effective output resistance of a practical Op Amp should be _____.

8. When the differential input voltage is zero, the Op Amp's output voltage ideally should be _____.

9. Feedback tends to _____ the effective output resistance and to _____ the effective input resistance.

10. Compared to a typical output signal voltage, the differential input should ideally be _____.

11. Comparatively, what do the terms *open-loop voltage gain* and *closed-loop voltage gain* mean referring to Op Amps?

12. To which of the Op Amp characteristics does the term *slew rate* apply?

13. If the Op Amp's output must be capable of swinging in both positive and negative directions, generally what kind of power supply does it require?

14. What does it mean when we say an Op Amp is *nulled*?

15. What is the meaning of the term *high CMRR*?

16. What undesirable effect might result if an Op Amp's dc source voltages are not well regulated?

17. A zener diode has a (*relatively constant*) (*varying*) voltage across it with a (*relatively constant*) (*varying*) current through it.

PROBLEMS

1. Design a half-wave zener-regulated power supply such as the circuit in Fig. 2-6a, with plus and minus 10-V dc outputs. Its loads—a few Op Amps and a relay—will draw currents varying from 10 mA to 50 mA. Select a transformer, zeners, capacitors, resistors, and rectifier diodes from the parts lists in Fig. 2-7. The transformer is to work off a 117-V line that can vary from 100 V to 130 V rms.

2. Design a full-wave zener-regulated power supply, such as the circuit of Fig. 2-6b, with plus and minus 12-V dc outputs. Its loads—a couple of Op Amps, relays, and transistor amplifiers—will draw currents varying from 15 mA to 100 mA. Select a transformer, zeners, capacitors, resistors, and rectifier diodes from the parts lists in Fig. 2-7. The transformer will work off a 117-V line that varies from 105 V to 125 V rms.

Wire-wound resistors

Ohms	3-W	5-W	10-W	Ohms	3-W	5-W	10-W
1	2701	2801A	2901A	100	2726	2826	2926
1.3	....	....	2901D	120	2727	2827	2927
1.5	2702	2802	2902	130	2727A	....	....
2	2703	2803	....	150	2728	2828	2928
2.2	2703A	....	....	160	2728A	2828A	....
2.4	....	2803B	....	180	2729	....	....
3	2704	2804	2904	200	2730	2830	2930
3.9	2704C	....	....	220	2731	2831	2931
4	2705	2805	....	225	2732	....	....
5	2706	2806	....	250	2733	2832	2932
5.6	2706B	....	....	270	2734	2833	....
6.2	....	....	2906C	300	2735	2834	2934
6.8	2706D	....	....	330	2736	2835	....
7.5	2707	2807	2907	350	2736A	2835A	....
10	2708	2808	2908	390	2737	....	....
12	....	2809	....	400	2738	2837	2937
15	....	2810	2910	430	2738A	....	....
18	....	2811	....	450	2738B	2837B	....
20	2712	2812	2912	470	2739	2838	....
22	2713	....	....	500	2740	2839	2939
25	2714	2814	2914	510	....	2839A	....
30	2716	2816	2916	560	2741	2840	....
33	2717	2817	....	600	2742	2841	....
35	....	....	2917A	620	2742A	....	2941A
39	2718	....	....	680	2743	2842	....
40	2719	2819	2919	750	2745	2844	....
47	2720	....	....	800	2746	2845	....
50	2721	2821	2921	820	2747	....	....
60	2723	....	....	1000	2749	2848	2948
68	2723B	....	....	1100	2749A	....	....
75	2724	2824	2924	1200	2750	2849	....
82	2725	2825	....	1300	2750A	....	....

Low-current
rectifier diodes

Type	PIV	Forward current	
		I_O Amps	$I_{FM(surge)}$ Amps
1N4001	50	1.0	30
1N4002	100	1.0	30
1N4003	200	1.0	30
1N4004	400	1.0	30
1N4005	600	1.0	30
1N4006	800	1.0	30
1N4007	1000	1.0	30

I_O is average forward current

Fig. 2-7 (Part 1)

Zener diodes
400 mW

Type	Nom. V_Z at I_{ZT} Volts	I_{ZT} mA	Max. Z_{ZT} at I_{ZT} Ω
1N957	6.8	18.5	4.5
1N958	7.5	16.5	5.5
1N959	8.2	15	6.5
1N960	9.1	14	7.5
1N961	10.0	12.5	8.5
1N962	11	11.5	9.5
1N963	12	10.5	11.5
1N964	13	9.5	13.0
1N965	15	8.5	16.0
1N966	16	7.8	17.0
1N967	18	7.0	21
1N968	20	6.2	25
1N969	22	5.6	29
1N970	24	5.2	33
1N971	27	4.6	41
1N972	30	4.2	49
1N973	33	3.8	58
1N974	36	3.4	70
1N975	39	3.2	80
1N976	43	3.0	93
1N977	47	2.7	105
1N978	51	2.5	125
1N979	56	2.2	150
1N980	62	2.0	185
1N981	68	1.8	230
1N982	75	1.7	270
1N983	82	1.5	330
1N984	91	1.4	400
1N985	100	1.3	500
1N986	110	1.1	750
1N987	120	1.0	900
1N988	130	0.95	1100
1N989	150	0.85	1500
1N990	160	0.80	1700
1N991	180	0.68	2200
1N992	200	0.65	2500

$1\frac{1}{2}$ watt zeners

Type	Nom. V_Z at I_{ZT} Volts	I_{ZT} mA	Max. Z_{ZT} at I_{ZT} Ω
1N3785	6.8	55	2.7
1N3786	7.5	50	3.0
1N3787	8.2	46	3.5
1N3788	9.1	41	4.0
1N3789	10	37	5
1N3790	11	34	6
1N3791	12	31	7
1N3792	13	29	8
1N3793	15	25	10
1N3794	16	23	11
1N3795	18	21	13
1N3796	20	19	15
1N3797	22	17	16
1N3798	24	16	17
1N3799	27	14	20
1N3800	30	12	25
1N3801	33	11	30
1N3802	36	10	35
1N3803	39	10	40
1N3804	43	9.0	45
1N3805	47	8.0	55
1N3806	51	7.4	65
1N3807	56	6.7	75
1N3808	62	6.0	85
1N3809	68	5.5	95
1N3810	75	5.0	110
1N3811	82	4.5	130
1N3812	91	4.1	150
1N3813	100	3.7	200
1N3814	110	3.4	300
1N3815	120	3.1	350
1N3816	130	2.9	400
1N3817	150	2.5	700
1N3818	160	2.3	750
1N3819	180	2.1	800
1N3820	200	1.9	1000

Power transformers:
Secondary voltages with
117 V on primary

No.	Secondary		VA rating
	Parallel	Series	
P-6375	6 V at 2 amps	12 V at 1 amp	12
P-6376	6 V at 4 amps	12 V at 2 amps	24
P-6377	12 V at 4 amps	24 V at 2 amps	48
P-6378	12 V at 8 amps	24 V at 4 amps	96
P-6379	12 V at 16 amps	24 V at 8 amps	192

Fig. 2-7 (Part 2)

Medium-current rectifier diodes

JEDEC type	PIV	Maximum I_{DC}‡ A	Maximum Peak surge	JEDEC type	PIV	Maximum I_{DC}‡ A	Maximum Peak surge
1N248C	55	20■	350 A	1N2155	100	25▲	400 A
1N249C	110	20■	350 A	1N2156	200	25▲	400 A
1N250C	220	20■	350 A	1N2157	300	25▲	400 A
1N1183	50	35●	500 A	1N2158	400	25▲	400 A
1N1184†	100	35●	500 A	1N2159	500	25▲	400 A
1N1185	150	35●	500 A	1N2160	600	25▲	400 A
1N1186†	200	35●	500 A	1N3208	50	15◆	250 A
1N1187	300	35●	500 A	1N3209	100	15◆	250 A
1N1188†	400	35●	500 A	1N3210	200	15◆	250 A
1N1189	500	35●	500 A	1N3211	300	15◆	250 A
1N1190	600	35●	500 A	1N3212	400	15◆	250 A
1N1195A	300	20■	350 A	1N3213	500	15◆	250 A
1N1196A	400	20■	350 A	1N3214	600	15◆	250 A
1N1197A	500	20■	350 A	1N3670A	700	12■	240 A
1N1198A	600	20■	350 A	1N3671A	800	12■	240 A
1N1199A	50	12■	240 A	1N3672A	900	12■	240 A
1N1200A	100	12■	240 A	1N3673A	1000	12■	240 A
1N1201A	150	12■	240 A	1N3765	700	35●	500 A
1N1202A	200	12■	240 A	1N3766	800	35●	500 A
1N1203A	300	12■	240 A	1N3767	900	35●	500 A
1N1204A	400	12■	240 A	1N3768	1000	35●	500 A
1N1205A	500	12■	240 A	1N5331	1200	12■	240A
1N1206A†	600	12■	240 A	1N5332	1200	35●	500 A
1N1341A	50	6■	150 A	A40A	100	20§	300 A
1N1342A	100	6■	150 A	A40C	200	20§	300 A
1N1343A	150	6■	150 A	A40G	300	20§	300 A
1N1344A	200	6■	150 A	A40D	400	20§	300 A
1N1345A	300	6■	150 A	A40E	500	20§	300 A
1N1346A	400	6■	150 A	A40F	50	20§	300 A
1N1347A	500	6■	150 A	A40M	600	20§	300 A
1N1348A	600	6■	150 A	A44A	100	20§	300 A
1N1612	50	5■	150 A	A44B	200	20§	300 A
1N1613	100	5■	150 A	A44C	300	20§	300 A
1N1614*	200	5■	150 A	A44D	400	20§	300 A
1N1615*	400	5■	150 A	A44E	500	20§	300 A
1N1016*	600	5■	150 A	A44F	50	20§	300 A
1N2135A*	400	20■	350 A	A44M	600	20§	300 A
1N2154	50	25▲	400 A				

I_{DC} is average forward current

Fig. 2-7 (Part 3)

5-watt Zeners

Type	Nom. V_Z at I_{ZT} Volts	I_{ZT} mA	Max. Z_{ZT} at I_{ZT} Ω
1N5339A	5.6	220	1.0
1N5341A	6.2	200	1.0
1N5342A	6.8	175	1.0
1N5343A	7.5	175	1.5
1N5344A	8.2	150	1.5
1N5346A	9.1	150	2.0
1N5347A	10	125	2.0
1N5348A	11	125	2.5
1N5349A	12	100	2.5
1N5350A	13	100	2.5
1N5352A	15	75	2.5
1N5353A	16	75	2.5
1N5355A	18	65	2.5
1N5357A	20	65	3.0
1N5358A	22	50	3.5
1N5359A	24	50	3.5
1N5361A	27	50	5.0
1N5363A	30	40	8.0
1N5364A	33	40	10
1N5365A	36	30	11
1N5366A	39	30	14
1N5367A	43	30	20
1N5368A	47	25	25
1N5369A	51	25	27
1N5370A	56	20	35
1N5372A	62	20	42
1N5373A	68	20	44
1N5374A	75	20	45
1N5375A	82	15	65
1N5377A	91	15	75
1N5378A	100	12	90
1N5379A	110	12	125
1N5380A	120	10	170
1N5381A	130	10	190
1N5383A	150	8	330
1N5384A	160	8	350
1N5386A	180	5	430
1N5388A	200	5	480

Fig. 2-7 (Part 4)

CHAPTER 3

THE OP AMP
WITH AND WITHOUT
FEEDBACK

The extremely high open-loop voltage gain A_{VOL} of the typical Op Amp makes it "touchy" to work with, especially in linear (undistorted) circuit applications. Because of the high gain, a relatively small differential input voltage V_{id} can easily drive an Op Amp's output voltage to its limit. If this occurs, the output signal is clipped and distorted. In this chapter we will see how negative feedback is used with an Op Amp to reduce and stabilize its effective voltage gain. In fact, with feedback we are able to select any specific gain we need as long as it is less than the Op Amp's open-loop gain A_{VOL}.

3-1 Open-Loop Considerations

The Op Amp is generally classified as a linear device. This means that its output voltage V_o tends to proportionally follow changes in the applied differential input V_{id}. Within limits, the changes in output voltage V_o are larger than the changes in the input V_{id} by the open-loop gain A_{VOL} of the Op Amp. The amount that the output voltage V_o can change (swing), however, is limited by the dc supply voltages and the load resistance R_L. Generally, the output voltage swing is restricted to values between the $+V$ and $-V$ supply voltages. As shown in Fig. 3-1, manufacturers provide curves showing their Op Amps' maximum output voltage swing versus supply voltage and also versus load resistance R_L. Note that smaller supply voltages and load resistances reduce an Op Amp's signal output capability. Any attempt to drive the Op Amp beyond its limits results in sharp clipping of its output signals.

39

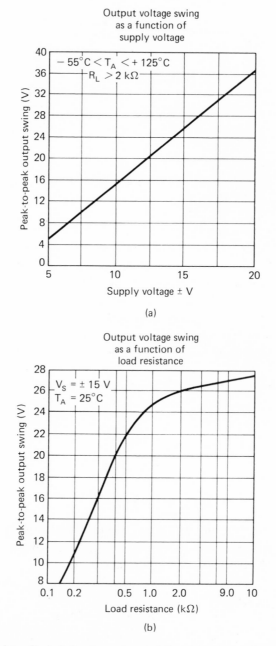

Fig. 3-1. Typical Op Amp characteristics: (a) output voltage V_o swing vs dc supply voltage; (b) output voltage V_o swing vs load resistance R_L.

For example, the Op Amp in Fig. 3-2 is shown with a small input signal V_{id} applied to its inverting input 1 and the resulting output V_o. Note that the output voltage $V_o = 0$ V at the times when $V_{id} = 0$ V, that is, the circuit is nulled. As we will see, a nulled output in a practical Op Amp is not easy to achieve if no feedback is used. Note also that the gain $A_{VOL} = 10,000$, which means that the output signal V_o is -10^4 times larger than the input V_{id}, but within limits. In this case, the limits are plus and minus 8 V, as indicated by the fact that the output signal V_o is clipped at $+8$ V on some positive alternations and at -8 V on some negative alternations. Apparently, this Op Amp has output voltage swing vs supply voltage characteristics as shown in Fig. 3-1. Note that, although the input V_{id} varies somewhat sinusoidally and with relatively low amplitudes, the output V_o is -10^4 times larger and easily reaches the positive and negative limits of the Op Amp. Consequently severe clipping occurs. Only the smaller input signals—1.6 mV peak to peak or less—are faithfully reproduced. There are applications where the Op Amp is purposely used as a signal-squaring circuit in which deliberately driving the Op Amp's input with a relatively large signal results in almost a square-wave output. This output appears more like a square and less like a trapezoid when the input signals have larger amplitudes and lower frequencies.

If the same input signal is applied to the noninverting input 2, as shown in Fig. 3-3, the output is $+10^4$ times larger, within limits of course. In this case, the output swings positively and then negatively on the positive and negative alternations of the input signal respectively. Clipping occurs when the output attempts to exceed the Op Amp's limits, just as in the inverting mode.

EXAMPLE 3-1

Sketch output voltage waveforms of an Op Amp, such as in Fig. 3-2, if the input signal is as shown but if:

(a) the open-loop gain $A_{VOL} = 5000$.
(b) the open-loop gain $A_{VOL} = 100,000$.

Answer. See Fig. 3-4. Note that with a larger gain, A_{VOL}, a given input signal tends to be more distorted. This does not mean that a high open-loop gain is undesirable, on the contrary, the larger A_{VOL} is the better. As we will see, clipping can easily be controlled with feedback.

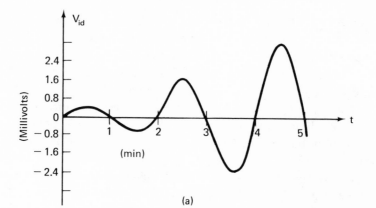

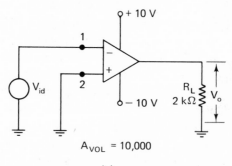

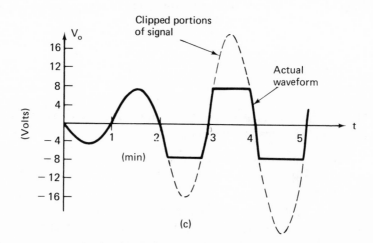

Fig. 3-2 (a) Differential input signal; (b) Op Amp with no feedback (inverting mode); (c) output signal waveform if $A_{VOL} = 10^4$.

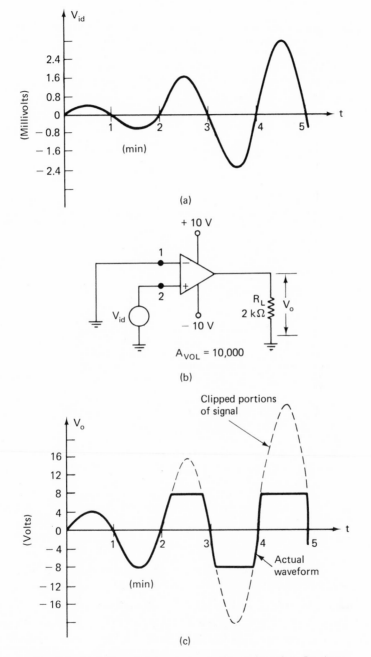

Fig. 3-3 (a) Differential input signal; (b) noninverting Op Amp circuit with no feedback; (c) output signal waveform.

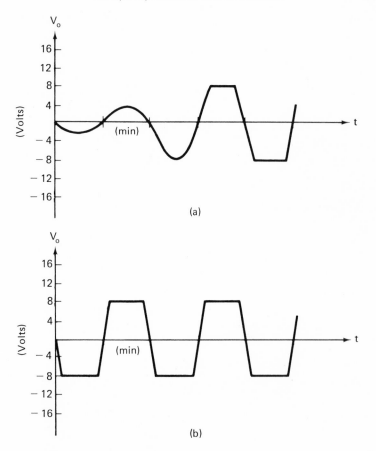

Fig. 3-4 (a) Output voltage of the Op Amp in Fig. 3-2 if $A_{VOL} = 5000$; (b) output voltage of the Op Amp in Fig. 3-2 if $A_{VOL} = 100,000$.

3-2   Feedback and the Inverting Amplifier

Some form of feedback is usually used with Op Amps in linear applications. Negative feedback enables the Op Amp circuit designer to easily select and control voltage gain. Generally, an amplifier has negative feedback if a portion of its output is fed back to its input and if the input and output signals are out of phase. The Op Amp circuit in Fig. 3-5 has negative feedback. Note that a resistor R_F is across the Op Amp's output and inverting input terminals. This, along with R_1, causes a portion of the output signal

V_o to be fed back to the inverting input terminal. With this type of feedback, the circuit's effective voltage gain A_v is typically much smaller than the Op Amp's open-loop gain A_{VOL}.

In the circuit shown in Fig. 3-5, the input signal is voltage V_s; therefore its effective voltage gain is

$$A_v = \frac{V_o}{V_s} \tag{3-1}$$

This ratio of output signal voltage V_o to the input signal V_s is also called the *closed-loop gain* because it is the gain when a feedback resistor R_F completes a loop from the amplifier's output to its inverting input 1. The specific value of A_v is determined mainly by the values of resistors R_1 and R_F.

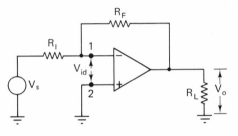

Fig. 3-5 Op Amp connected to work as an inverting amplifier with feedback.

We can see how resistors R_1 and R_F in the circuit of Fig. 3-5 determine the closed-loop gain A_v if we analyze this circuit's current paths and voltage drops. For example, as shown in Fig. 3-6, the signal source V_s drives a current I through R_1. Assuming that the Op Amp is ideal (having infinite resistance looking into input 1), all of this current I flows up through R_F. By Ohm's law we can show that the voltage drop across R_1 is $R_1 I$ and that the voltage across R_F is $R_F I$. The significance of this will be seen shortly.

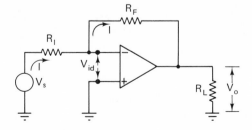

Fig. 3-6 Currents in R_1 and R_F are equal if the Op Amp is ideal; they are approximately equal with practical Op Amp if R_F is not too large.

Since the open-loop gain A_{VOL} of the Op Amp is ideally infinite or at least very large practically, the differential input voltage V_{id} is infinitesimal compared to the output voltage V_o. Thus since

$$A_{VOL} = \frac{V_o}{V_{id}} \qquad (2\text{-}3a)$$

then

$$V_{id} = \frac{V_o}{A_{VOL}}$$

This last equation shows that, the larger A_{VOL} is with any given V_o, the smaller V_{id} must be compared to V_o. The point is, the voltage V_{id} across inputs 1 and 2 is practically zero because of the large open-loop gain A_{VOL} of the typical Op Amp. Therefore, with *virtually* no potential difference between inputs 1 and 2, and with the noninverting input 2 grounded, the inverting input is *virtually* grounded too. This circuit's equivalent can therefore be shown as in Fig. 3-7. With voltage V_o to ground at the right of R_F and virtual ground at its left, the voltage across R_F is V_o for practical purposes. By Ohm's law

$$V_o \cong R_F I \qquad (3\text{-}2)$$

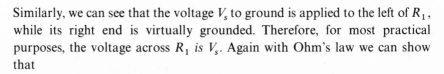

Inverting input
is virtually
grounded

Fig. 3-7 Equivalent circuit of Figs. 3-5 and 3-6; $V_s \cong R_1 I$ and $V_o \cong R_F I$.

Similarly, we can see that the voltage V_s to ground is applied to the left of R_1, while its right end is virtually grounded. Therefore, for most practical purposes, the voltage across R_1 is V_s. Again with Ohm's law we can show that

$$V_s \cong R_1 I \qquad (3\text{-}3)$$

Since the closed-loop gain A_v is the ratio of the output signal voltage V_o to the input signal voltage V_s, we can substitute Eqs. (3-2) and (3-3) into this ratio and show that

$$A_v = \frac{V_o}{V_s} \simeq -\frac{R_F I}{R_1 I} \simeq -\frac{R_F}{R_1} \qquad (3\text{-}4)$$

The negative sign means that the input and output signals are out of phase. This last equation shows that by selecting a ratio of feedback resistance R_F to the input resistance R_1, we select the inverting amplifier's closed-loop gain A_v.

There are limits on the usable values of R_F and R_1 which are caused by practical design problems. Though these design problems are discussed in later chapters, for the present we should know that the feedback resistance R_F is rarely larger than 10 MΩ. More frequently, R_F is 1 MΩ or less. Since one end of the signal source V_s is grounded and one end of R_1 is virtually grounded, V_s sees R_1 as the amplifier's input resistance. To avoid loading of (excessive current drain from) the signal source V_s, R_1 is typically 1 kΩ or more.

EXAMPLE

Referring to the circuit in Fig. 3-8a, find its voltage gain V_o/V_s when the switch S is in:
(a) position 1,
(b) position 2, and
(c) position 3.
(d) If the switch is in the 2 position and the input voltage V_s has the waveform shown in Fig. 3-8b, sketch the output voltage waveform V_o onto Fig. 3-8c. Assume that when V_s was zero, the output V_o was zero (nulled).
(e) What is the resistance seen by the signal source V_s for each of the switch positions.

Answer. (a) When the switch S is in position 1, the feedback resistance $R_F = 10$ kΩ. The input resistor $R_1 = 1$ kΩ regardless of the switch position. Therefore, the gain is

$$A_v \simeq -\frac{R_F}{R_1} = -\frac{10 \text{ k}\Omega}{1 \text{ k}\Omega} = -10 \qquad (3\text{-}4)$$

(b) With the switch S in position 2, $R_F = 100$ kΩ; therefore

$$A_v \simeq -\frac{100 \text{ k}\Omega}{1 \text{ k}\Omega} = -100$$

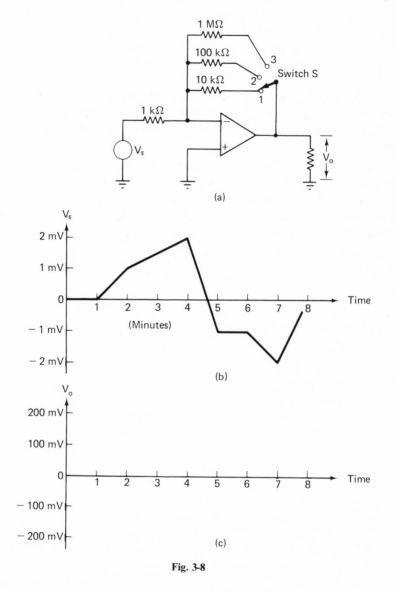

Fig. 3-8

(c) With the switch S in position 3, $R_F = 1$ MΩ; therefore

$$A_v \cong -\frac{1 \text{ M}\Omega}{1 \text{ k}\Omega} = -1000$$

The negative signs mean that the input and output signal voltages are out of phase.

(d) Since the gain $A_v = -100$ with the switch S in position 2, the input signal voltage V_o is amplified by 100 and its phase is inverted, resulting in an output voltage waveform shown in Fig. 3-9. By rearranging Eq. (3-1), we can show that

$$V_o = A_v V_s$$

Thus at $t = 2$ min, $V_o = -100$ (1 mV) $= -100$ mV. At $t = 4$ min, $V_o = -100$ (2 mV) $= -200$ mV, etc.

(e) The signal source V_s sees R_1 as the load regardless of the switch position.

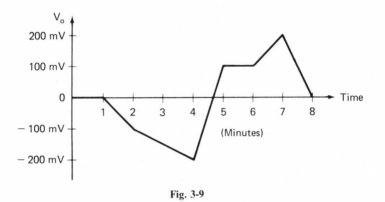

Fig. 3-9

EXAMPLE 3-3

The Op Amp in Fig. 3-8 operates with dc supply voltages of ± 15 V and with a load resistance $R_L = 200\ \Omega$. If the characteristics shown in Fig. 3-1 are typical of this Op Amp, what are the values of:

   (a) the maximum possible peak-to-peak unclipped output signal V_o, and
   (b) the maximum input peak-to-peak signal voltage V_s that can be applied and not cause clipping of V_o while the switch S is in the 2 position?

Answer. (a) Referring to Fig. 3-1b, we project up from 0.2 kΩ to the curve. Directly to the left of the intersection of our projection and curve, we see that this amplifier's peak-to-peak output swing capability is about 11 V. This means that clipping will occur if we attempt to drive the output V_o beyond $+5.5$ V or -5.5 V.

(b) With the switch S in position 2, the gain is -100 as we found in the previous problem. Since the maximum peak-to-peak output swing is about 11 V, the maximum peak-to-peak input swing is determined as follows. Since

$$A_v = \frac{V_o}{V_s} \qquad (3\text{-}1)$$

then

$$V_s = \frac{V_o}{A_v} \cong \frac{11\,\text{V (p-p)}}{-100} = |110\,\text{mV (p-p)}|$$

3-3   The Noninverting Amplifier

We can use the Op Amp, with feedback, in a noninverting mode much as we used it in the inverting mode discussed in the previous section. We can drive the noninverting input 2 with a signal source V_s instead of indirectly driving the inverting input 1 as shown in Figs. 3-3 and 3-10. As with the inverting amplifier, the values of externally connected resistors R_1 and R_F determine the circuit's closed-loop voltage gain A_v. A gain equation in terms of R_1 and R_F can be worked out if we analyze the noninverting amplifier's currents and voltages.

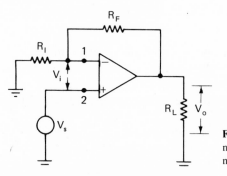

Fig. 3-10 Op Amp wired to work as a noninverting amplifier (noninverting mode).

As shown in Fig. 3-11a, the signal current I is the same through resistors R_1 and R_F as long as the resistance R_i looking into the inverting input 1 is infinite or at least very large. In other words, R_1 and R_F are effectively in series. As before, the voltage drops across these resistors can be shown as R_1I and R_FI. At the right side of R_F we have the output voltage V_o to ground.

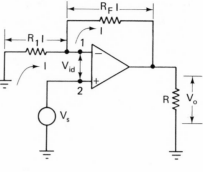

(a)

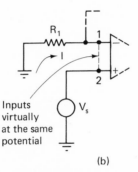

Fig. 3-11 (a) Noninverting amplifier with currents and voltage drops shown; (b) the difference voltage V_{id} across inputs 1 and 2 is so small compared to V_s that these inputs are virtually shorted.

Inputs virtually at the same potential

(b)

This voltage is also across R_1 and R_F because these resistors are effectively in series and the left side of R_1 is grounded. Thus we can show that the output voltage V_o is the sum of the drops across R_1 and R_F, that is

$$V_o \cong R_1 I + R_F I$$

Factoring I out, we get

$$V_o \cong (R_1 + R_F)I \qquad (3\text{-}5)$$

As with the inverting amplifier, the differential input voltage V_{id} is zero for most practical purposes. Due to the very large open-loop gain A_{VOL} of the typical Op Amp, we can assume that there is practically no potential difference between points 1 and 2 as shown in Fig. 3-11b. Therefore, nearly all of the input signal voltage V_s appears across R_1, and by Ohm's law, we can still show that

$$V_s \cong R_1 I \qquad (3\text{-}3)$$

The voltage gain A_v of the noninverting amplifier is the ratio of its output V_o to its input V_s. Thus substituting the right sides of Eqs. (3-3) and (3-5) into this ratio yields

$$A_v = \frac{V_o}{V_s} \cong \frac{(R_F + R_1)I}{R_1 I} = \frac{R_F + R_1}{R_1} = \frac{R_F}{R_1} + 1 \qquad (3\text{-}6)$$

Since the resistance seen looking into the noninverting input is infinite if the Op Amp is ideal (or at least very large with a practical Op Amp), the signal source V_s sees a very large resistance looking into the Op Amp of Fig. 3-10. Generally, the effective input resistance of the noninverting amplifier is larger with circuits wired to have lower closed-loop gains. More specifically, the effective opposition to signal current flow into the non-inverting input is an impedance $Z_{i(eff)}$ instead of a resistance when input capacitances are considered. Looking into the noninverting input 2, the signal current sees parallel paths: through an impedance Z_{ic} and through the Op Amp's resistance R_i. See Fig. 3-12. Thus the total effective input impedance is

$$Z_{i(eff)} \cong \frac{1}{A_v/A_{VOL}R_i + 1/Z_{ic}} \qquad (3\text{-}7a)$$

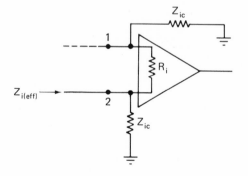

Fig. 3-12 Equivalent circuit of an Op Amp; R_i is the resistance between inputs 1 and 2 specified by the manufacturer; Z_{ic} is the impedance to ground or common from either input which is actually internal to the Op Amp.

where: R_i is the input resistance measured between inputs 1 and 2, open-loop,

Z_{ic} is the impedance to ground or common measured from either input,

A_{VOL} is the open-loop gain, and

A_v is the closed-loop gain.

Typically, Z_{ic} is much larger than R_i, especially at low frequencies. If we assume that Z_{ic} is relatively very large, Eq. (3-7a) can be simplified to

$$Z_{i(\text{eff})} \cong \frac{1}{A_v/A_{VOL}R_i} = \left(\frac{A_{VOL}}{A_v}\right)R_i \qquad (3\text{-}7b)$$

where the ratio A_{VOL}/A_v is often called the *loop gain*. Note that if the Op Amp is wired to have a low closed-loop gain A_v, say approaching 1, the effective input impedance $Z_{i(\text{eff})}$ theoretically approaches a value that is A_{VOL} times larger than the manufacturer's specified R_i. In such cases the shunt impedance to ground Z_{ic} would not be relatively large and would not be ignored if accuracy were important. Therefore, at low frequencies, the lower the closed-loop gain, the larger the effective input impedance $Z_{i(\text{eff})}$ of the noninverting amplifier.

EXAMPLE 3-4

If $R_1 = 1\,\text{k}\Omega$ and $R_F = 10\,\text{k}\Omega$ in the circuit of Fig. 3-10, what is the voltage gain V_o/V_s? If this circuit's input signal voltage V_s has the waveform shown in Fig. 3-13a, sketch the output waveform V_o onto Fig. 3-13b. Assume that the Op Amp's output was initially nulled.

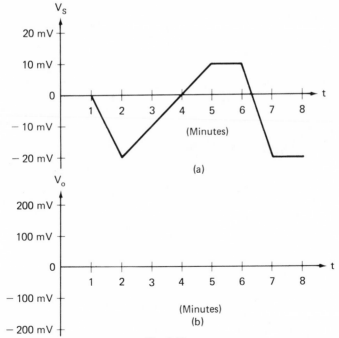

Fig. 3-13

Answer. Since this Op Amp is wired in a noninverting mode, the voltage gain can be determined with Eq. (3-6):

$$A_v \cong \frac{R_F}{R_1} + 1 = 11$$

Therefore, the output voltage is 11 times larger than, and in phase with, the input signal V_s. Thus at $t = 2$ minutes, $V_o = 11(-20\,\text{mV}) = -220\,\text{mV}$. Similarly, at $t = 5$ minutes, $V_o = 11(10\,\text{mV}) = 110\,\text{mV}$, etc. See Fig. 3-14 for the complete output waveform.

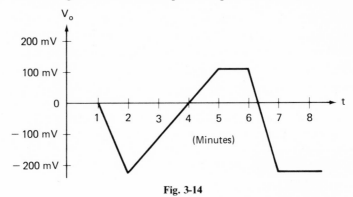

Fig. 3-14

EXAMPLE 3-5

Refer to the circuit described in Example 3-4. If the Op Amp is a 741C type as listed in Appendix SI, what approximate minimum impedance does the signal source V_s see if the impedance to ground from either input is assumed to be infinite? Assume that Z_{ic} is infinite.

Answer. As shown in Appendix SI, the 741C's minimum A_{VOL} and R_i values are 20,000 and 150 kΩ, respectively. Since the circuit is wired externally for a closed-loop gain of 11, we can find the minimum effective impedance seen by V_s with Eq. (3-7b). Thus in this case

$$Z_{i(\text{eff})} \cong \left(\frac{20,000}{11}\right) 150\,\text{k}\Omega \cong 273\,\text{M}\Omega \qquad (3\text{-}7b)$$

3-4   The Voltage Follower

In some applications, an amplifier's voltage gain is not as important as its ability to match a high internal resistance signal source to a low, possibly

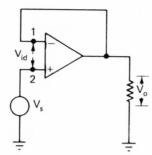

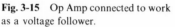

Fig. 3-15 Op Amp connected to work as a voltage follower.

varying, resistance load. The Op Amp in Fig. 3-15 is connected to work as a *voltage follower* which has an extremely large input resistance and is capable of driving a relatively low-resistance load. The voltage follower's output resistance is very small; therefore, variations in its load resistance negligibly affect the amplitude of the output signal. When used between a high-internal-resistance signal source and a smaller, varying-resistance load, the voltage follower is called a *buffer amplifier*.

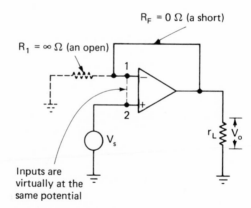

Fig. 3-16 The voltage follower of Fig. 3-10 can be shown to be a non-inverting amplifier.

Inputs are virtually at the same potential

The voltage follower is simply a noninverting amplifier, similar to the one in Fig. 3-11a, where R_1 is replaced with infinite ohms (an open) and R_F is replaced with zero ohms (a short). See Fig. 3-16. Due to the Op Amp's large open-loop gain A_{VOL}, the differential input voltage V_{id} is very small; therefore, the inputs 1 and 2 are virtually at the same potential. Since the output signal V_o is the voltage at input 1, and since the input signal V_s is directly applied to input 2, then

$$V_o \cong V_s$$

Apparently, with the input and output signal voltages nearly equal, the voltage gain of the voltage follower is very nearly 1 (unity). We can see this another way—by substituting $R_F = 0\,\Omega$ into the gain Eq. (3-6). The term *voltage follower* therefore describes the circuit's function: the output voltage V_o follows the input voltage V_s waveform.

As with the noninverting amplifier, the effective input impedance $Z_{in(eff)}$ of the voltage follower is about the loop gain times larger than R_i, where R_i is the differential input resistance measured under open-loop conditions. Since the voltage follower's closed-loop gain $A_v \cong 1$, its loop gain A_{VOL}/A_v is very large—about equal to A_{VOL}. This accounts for the extremely high input impedance of the voltage follower.

As the input impedance $Z_{in(eff)}$ is made more ideal (increased) by the use of negative feedback, the output resistance is also improved (reduced). More specifically, the effective output resistance $R_{o(eff)}$ of an Op Amp is smaller than the output resistance R_o measured with open-loop conditions by the loop-gain factor. That is,

$$R_{o(eff)} \cong \left(\frac{A_v}{A_{VOL}}\right)R_o \qquad (3\text{-}8)$$

This shows that, the smaller the closed-loop gain A_v, the smaller the effective output resistance $R_{o(eff)}$.

EXAMPLE 3-6

   (a) If the circuit in Fig. 3-15 has the input voltage waveform V_s shown in Fig. 3-9, sketch the output voltage waveform. Assume that initially the Op Amp was nulled.
   (b) If a signal source has 100 kΩ of internal resistance and an output of 4 mV *before* being connected to input 2 (open circuited), what is the output of the signal source *after* it is connected to the input 2 of this circuit?

Answer. (a) Since the gain A_v of the voltage follower is unity, and since there is no phase inversion, the output V_o has the same waveform as the input V_s.

(b) The input impedance of the voltage follower is in the range of hundreds of megohms and therefore appears as an open compared to the 100 kΩ internal resistance of the signal source. Thus there is essentially no signal voltage drop across the internal resistance, and the output of the signal generator remains very close to its 4-mV open-circuit value after being connected to input 2. Therefore the output of this voltage follower is also about 4 mV.

REVIEW QUESTIONS

1. What two factors outside of the Op Amp affect its maximum unclipped output signal capability?

2. When the Op Amp is connected as an inverting amplifier (Fig. 3-5), what is the approximate input resistance as seen by the signal source V_s?

3. Since the Op Amp's open-loop gain is usually very high, why is it that the Op Amp is seldom used as a signal amplifier in open loop?

4. If a low-frequency sine wave with a 1-V peak-to-peak amplitude were directly applied to an Op Amp's inputs 1 and 2, what kind of waveform would you expect at the output?

5. What is the difference between the closed-loop gain and the open-loop gain?

6. How does the differential input signal V_{id} compare in magnitude to the output voltage V_o?

7. How can an Op Amp input terminal be virtually grounded but not actually grounded?

8. How does the input resistance of a noninverting amplifier compare with the input resistance of the inverting type?

9. If an Op Amp is described as having been nulled, what does this mean?

10. What effect does negative feedback have on the voltage gain of an amplifier?

11. If an Op Amp is connected to have negative feedback, how does its effective output resistance compare with the output resistance specified by the manufacturer?

12. If an Op Amp is connected to operate as a noninverting amplifier, with a closed-loop gain lower than the open-loop gain, how does its effective input impedance compare with the input resistance specified by the manufacturer?

13. Which would you expect to have the larger input impedance: an Op Amp connected to work as a noninverting amplifier with a closed-loop gain of 101, or the same Op Amp connected to work as a voltage follower?

14. Since the voltage follower has only unity voltage gain, what good is it?

15. What might happen to the output signal voltage waveform if the load resistance is too small?

16. What might happen to the output signal voltage waveform if the dc supply voltages are too small?

PROBLEMS

In the circuit of Fig. 3-17a, the signal driving the Op Amp is taken off a bridge circuit as shown. The thermistor in the bridge is a resistance with a

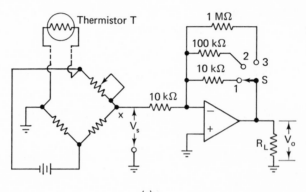

(a)

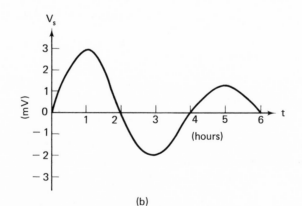

(b)

Fig. 3-17

high negative temperature coefficient, that is, its resistance significantly decreases or increases as its temperature increases or decreases, respectively. The thermistor is in an oven. As the oven's temperature rises or falls, the voltage V_s swings positively or negatively. The load R_L represents the effective resistance of a motor's armature. When the shaft of the motor turns, it operates a fuel valve and thus admits more or less fuel to the oven. This system therefore controls the temperature in the oven. The switch S provides a means of selecting its sensitivity (closed loop voltage gain).

If the voltage across the bridge, which is voltage V_s, has the waveform shown in Fig. 3-17b, find the voltage gain A_v and select a waveform from

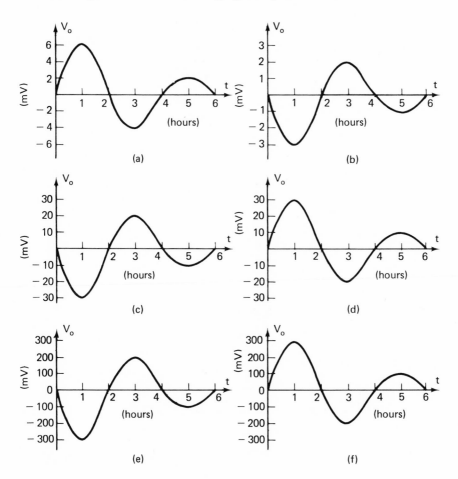

Fig. 3-18

Fig. 3-18 that best represents the Op Amp's output V_o for each of the following conditions:

1.   Switch S is in position 1.

2.   Switch S in in position 2.

3.   Switch S is in position 3.

4.   What is the load resistance across the bridge (points x and ground) in the circuit of Fig. 3-17a?

If the Op Amp is removed from the system and replaced with the one in Fig. 3-19, but the waveform V_s is still as shown in Fig. 3-17b, find this Op

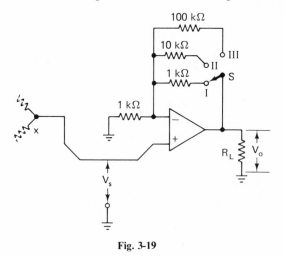

Fig. 3-19

Amp's voltage gain and select a waveform from Fig. 3-18 that best represents the Op Amp's output V_o for each of the following conditions:

5.   The switch S is in the I position.

6.   The switch S is in the II position.

7.   The switch S is in the III position.

8.   Which of the Op Amps—the one in Fig. 3-17a or the one in Fig. 3-19— draws the larger current from the bridge?

9.   Which of the Op Amps—the one in Fig. 3-17a or the one in Fig. 3-19— offers the more constant load on the bridge (keep in mind that the switch S position might frequently be changed)?

The switches S_1 and S_2 in the circuit of Fig. 3-20 are ganged, that is, when S_1 is in the 1 position, S_2 is also in the 1 position. Similarly, S_1 and S_2 are in the 2 position simultaneously.

10. If the signal source is 2 V dc, $R_s = 50\,k\Omega$, and $R_L = 1\,k\Omega$ in the circuit of Fig. 3-20, what is the approximate voltage at point 0 to ground when S_1 is in the 2 position?

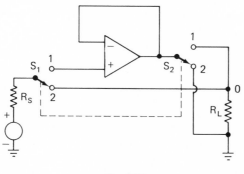

Fig. 3-20

11. If the signal source is 3 V dc, $R_s = 150\,k\Omega$, and $R_L = 1\,k\Omega$ in the circuit of Fig. 3-20, what is the approximate voltage at point 0 with respect to ground when S_1 is in the 2 position?

12. Refer to Problem 10. What is the voltage at point 0 with respect to ground when the switches are in the 1 position?

13. Refer to Problem 11. What is the voltage at point 0 with respect to ground when the switches are in the 1 position?

14. What is the typical effective output resistance of the 709 Op Amp whose characteristics are given in Appendix SII, when it is connected to work as a voltage follower?

15. What is the typical effective output resistance of the 741 Op Amp whose characteristics are given in Appendix SIII, when it is connected to work with a closed-loop gain A_v of 1000?

16. Over what range can the gain A_v of the circuit shown in Fig. 3-21a be adjusted if the resistor R can be varied from $0\,\Omega$ to $1\,M\Omega$?

17. If $R = 190\,k\Omega$ in the circuit of Fig. 3-21a, what is the output voltage V_o if $V_s = -2\,mV$ dc?

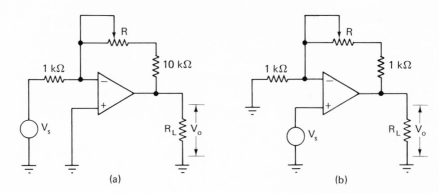

(a)   (b)

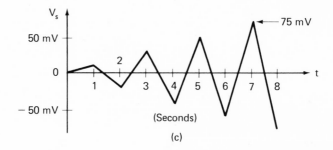

(c)

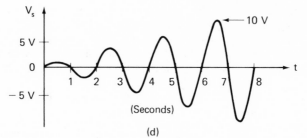

(d)

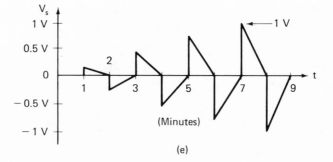

(e)

Fig. 3-21

18. If $V_s = -3$ mV dc and $R = 5$ kΩ in the circuit shown in Fig. 3-21b, what is the output voltage V_o?

19. If resistance R can be adjusted from 0 Ω to 200 kΩ in the circuit of Fig. 3-21b, over what range can its voltage gain A_v be varied?

The following four problems refer to the circuit in Fig. 3-21a with the input voltage waveform as shown in Fig. 3-21c. The Op Amp was initially nulled and has the characteristics in Fig. 3-1.

20. Sketch the output waveform V_o if the dc supply voltages are +15 V and −15 V, the resistance R is adjusted to 90 kΩ, and $R_L = 2$ kΩ.

21. Sketch the output waveform V_o if the dc supply voltages are ±15 V, the resistance R is adjusted to 140 kΩ, and $R_L = 1$ kΩ.

22. Referring to Problem 20, sketch the output waveform V_o if the dc supply voltages are reduced from ±15 V to ±7.5 V. The R and R_L values are unchanged.

23. Referring to Problem 21, sketch the output waveform V_o if the load resistance R_L is reduced to 200 Ω.

The following three problems refer to the circuit in Fig. 3-21b with the input voltage as shown in Fig. 3-21d. The Op Amp was initially nulled and has the characteristics in Fig. 3-1.

24. Sketch the output waveform V_o if the dc supply voltages are ±15 V, the resistance R is adjusted to 0 Ω and $R_L = 2$ kΩ.

25. Sketch the output waveform V_o if the dc supply voltages are ±5 V, the resistance R is adjusted to 0 Ω, and $R_L = 2$ kΩ.

26. Sketch the output waveform V_0 if the dc supply voltages are ±15 V, the resistance R is adjusted to 0 Ω, and $R_L = 200$ Ω.

27. Refer to Fig. 3-21a. If the circuit has the input waveform shown in Fig. 3-21e, $R = 40$ kΩ, $R_L = 2$ kΩ, ±15-V supply voltages, and the Op Amp characteristics in Fig. 3-1, show the output voltage waveform.

28. Rework the previous problem using the circuit in Fig. 3-21b instead. The values of R, R_L, and supply voltages are unchanged.

CHAPTER 4

OFFSET
CONSIDERATIONS

The practical Op Amp, unlike the hypothetical ideal version, has some dc output voltage, called *output offset* voltage, even though both of its inputs are grounded. Such an output offset is an error voltage and is generally undesirable. The causes and cures of output offset voltages are the subjects of this chapter. Here we will become familiar with parameters that enable us to predict the maximum output offset voltage that a given Op Amp circuit can have. On the foundation laid in this chapter, we will build an understanding of why, and an ability to predict how much, a given Op Amp's output voltage tends to drift with power supply and temperature changes which are important subjects discussed later.

4-1   Input Offset Voltage V_{io}

The input offset voltage V_{io} is defined as the amount of voltage required across an Op Amp's inputs 1 and 2 to force the output voltage to 0 V. In a previous chapter we learned that ideally, when the differential input voltage $V_{id} = 0$ V, as in Fig. 4-1, the output offset V_{oo} is 0 V too. In the practical case, however, some V_{oo} voltage will usually be present. This is caused by imbalances within the Op Amp's circuitry. They occur because the transistors of the input differential stage within the Op Amp usually admit slightly

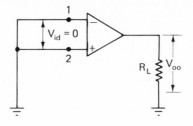

Fig. 4-1   Ideally, $V_{oo} = 0$.

different collector currents even though both bases are at the same potential. This causes a differential output voltage from the first stage, which is amplified and possibly aggravated by more imbalances in the following stages. The combined result of these imbalances is the output offset voltage V_{oo}. If small dc voltage of proper polarity is applied to inputs 1 and 2, it will decrease the output voltage. The amount of differential input voltage, of the correct polarity, required to reduce the output to 0 V is the input offset voltage V_{io}. When the output is forced to 0 V by the proper amount and polarity of input voltage, the circuit is said to be *nulled* or *balanced*. The polarity of the required input offset voltage V_{io} at input 1 with respect to input 2 might be positive as often as negative. This means that the output offset voltage, before nulling, can be positive or negative with respect to ground.

Common nulling circuits of inverting-mode amplifiers are shown in Fig. 4-2. Their equivalents for noninverting amplifiers are shown in Fig. 4-3.

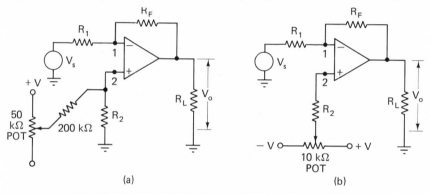

Fig. 4-2 Typical null balancing circuits in inverting amplifiers.

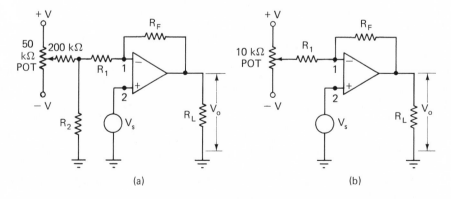

Fig. 4-3 Typical null balancing circuits in noninverting amplifiers.

In each of these circuits, the potentiometer POT can be adjusted to provide the value and polarity of dc input voltage required to null the output to 0 V. As shown, the ends of the POT are connected to the positive and negative dc supply voltages. With some Op Amps, such as types 748, 777, 201, 741,* etc. (see Appendix F1), *offset adjust* pins are provided. The manufacturers recommend that a POT be placed across their OFFSET NULL pins as shown in Fig. 4-4. Adjustment of this POT will null the output if the closed loop gain A_v is not too large.

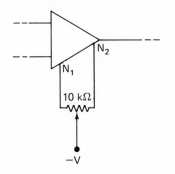

Fig. 4-4 Null adjustment on an Op Amp type that has null adjustment pins provided.

If an Op Amp circuit is not nulled, more or less output offset voltage V_{oo} exists, depending on its specified input offset voltage V_{io}, closed-loop gain A_v, and other factors discussed in the following sections. When the source voltage $V_s = 0$ V, an equivalent circuit of *either* the inverting *or* the non-inverting amplifier is as shown in Fig. 4-5. In the figure, the Op Amp's

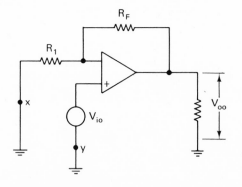

Fig. 4-5 Equivalent circuit of either the inverting or the noninverting amplifier.

---

* Generally, 741-type Op Amps have characteristics such as listed in Appendix SIII. 741s are identified in several ways, depending on the manufacturer. To name a few: Fairchild μA741, Motorola MC1741, National Semiconductor LM741, Raytheon RM741, Signetics, μA741.

specified input offset voltage V_{io} is shown as a dc signal source working into a noninverting type amplifier. If this equivalent circuit in Fig. 4-5 represents an inverting amplifier, the signal source V_s is replaced with its own internal resistance at point x. If this equivalent circuit represents a noninverting amplifier, V_s is replaced with its own internal resistance at point y. A short circuit replaces V_s if its internal resistance is negligible. According to Fig. 4-5 then, we can show that the output offset voltage V_{oo}, "caused by" the input offset voltage, is the product of the closed-loop gain and the specified V_{io}:

$$V_{oo} = A_v V_{io} \qquad (4\text{-}1)$$

where

$$A_v \cong \frac{R_F}{R_1} + 1 \qquad (3\text{-}6)$$

EXAMPLE

If the Op Amp in the circuit of Fig. 4-6 is a 741 type (see Appendix SIII), what is the *maximum* possible output offset voltage V_{oo}, caused by the input offset voltage V_{io}, before any attempt is made to null the circuit with the 10 kΩ POT?

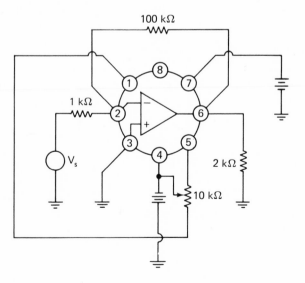

Fig. 4-6

Answer. According to the specifications of the 741, its maximum $V_{io} = 6$ mV. Since the input resistance $R_1 = 1$ kΩ, assuming that the internal resistance of V_s is negligible and since the feedback resistance $R_F = 100$ kΩ, the input offset voltage is multiplied by

$$A_v \cong \frac{R_F}{R_1} + 1 = \frac{100 \text{ k}\Omega}{1 \text{ k}\Omega} + 1 = 101$$

Therefore, before the circuit is nulled, we might have an output offset as large as

$$V_{oo} = A_v V_{io} \cong 101 (6 \text{ mV}) = 606 \text{ mV} \quad \text{or about } 0.6 \text{ V}$$

This means that the output voltage with respect to ground can be either a positive or a negative 0.6 V even though the input signal $V_s = 0$ V.

4-2 Input Bias Current

Most types of IC Op Amps have two transistors in the first (input) differential stage. Transistors, being current-operated devices, require some base bias currents. Therefore, small dc bias currents flow in the input leads of the typical Op Amp as shown in Fig. 4-7. An input bias current I_B is usually specified on the Op Amp specification sheets as shown in the Appendices.

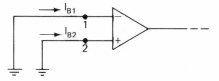

Fig. 4-7 Base bias currents flow into the Op Amp and return to ground through the power supplies.

It is defined as the average of the two base bias currents, that is,

$$I_B = \frac{I_{B_1} + I_{B_2}}{2} \tag{4-2}$$

These base bias currents, I_{B_1} and I_{B_2}, are *about* equal to each other, and therefore the specified input bias current I_B is *about* equal to either one of them, that is,

$$I_B \cong I_{B_1} \cong I_{B_2} \tag{4-3}$$

Depending on the type of Op Amp, the value of input bias current is usually small, generally in the range from a few to a few hundred nanoamperes.

Though this seems insignificant, it can be a problem in circuits using relatively large feedback resistors, as we will see.

A difficulty that the base bias currents might cause can be seen if we analyze their effect on a typical amplifier such as in Fig. 4-8. Note that I_{B_2} flows out of ground directly into input 2. (See Fig. 1-8 for the current paths of a typical differential input stage.) Therefore input 2 is 0 V with respect to ground. On the other hand, base bias current I_{B_1} sees resistance on its way to input 1, that is, the parallel paths containing R_1 and R_F have a total effective resistance as seen by the base bias current I_{B_1}. The flow of I_{B_1} through this effective resistance causes a dc voltage to appear at input 1.

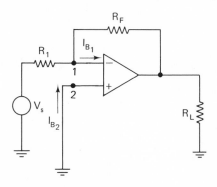

Fig. 4-8 Current I_{B_1} sees some resistance on its way to ground, whereas I_{B_2} sees a short.

With a dc voltage to ground at input 1 while input 2 is 0 V to ground, a dc differential input voltage V_{id} appears *across* these inputs. This dc input amplified by the closed-loop gain causes an output offset voltage V_{oo}. The amount of output offset voltage V_{oo}, caused by the base bias current I_{B_1}, can be approximated with the equation*

$$V_{oo} \cong R_F I_B \qquad (4\text{-}4)$$

where $I_B \cong I_{B_1}$, and is usually specified by the manufacturer.

EXAMPLE 4-2

Referring to the circuit in Fig. 4-6, what maximum output offset voltage might it have, caused by the base bias current, before it is nulled? The Op Amp is a type 741.

Answer. According to the 741's specification sheets (Appendix SIII), the maximum input bias current $I_B = 500\,\text{nA}$. Since $R_F = 100\,\text{k}\Omega$,

---

* See Appendix A1 for the derivation.

the base bias current could cause an output offset as large as

$$V_{oo} \cong R_F I_B = 100 \text{ k}\Omega \,(500 \text{ nA}) = 50 \text{ mV} \qquad (4\text{-}4)$$

We should note in these last two examples that the output offset caused by the input offset voltage V_{io} is about ten times larger than the output offset caused by the input bias current I_B. In this case then, the input offset voltage V_{io} is potentially the greater problem. However, according to Eq. (4-4), we can see that the output offset V_{oo} caused by the bias current I_B is larger, and therefore potentially more troublesome, with larger values of feedback resistance R_F.

EXAMPLE 4-3

Referring again to the circuit of Fig. 4-6, suppose that we replace the 1 kΩ input resistor with 100 kΩ and the 100 kΩ feedback resistor with 10 MΩ:

(a) How does this affect the closed-loop gain A_v of this circuit compared to what it was originally? Before any attempt is made to null this circuit, find

(b) Its maximum output offset caused by the input offset voltage V_{io}, and

(c) The maximum output offset caused by the input bias current I_B? The Op Amp is still a type 741.

Answer. (a) The closed-loop gain A_v is unchanged, that is, this circuit's gain is very nearly equal to the ratio R_F/R_1 which was not changed.

(b) With no change in the closed-loop gain, the output offset caused by the input offset voltage does not change. See Eq. (4-1).

(c) The output offset caused by input bias current I_B is larger with circuits using larger feedback resistors, according to Eq. (4-4). Thus in this case

$$V_{oo} \cong R_F I_B = 10 \text{ M}\Omega\,(500 \text{ nA}) = 5 \text{ V}$$

This is a relatively large output offset. It approaches the maximum swing capability of some Op Amp circuits, especially if fairly small dc supply voltages or load resistance R_L are used. This example presents a good case of the use of small feedback resistors.

The effect of the input bias current I_B on the output offset voltage can be minimized if a resistor R_2 is added in series with the noninverting input as shown in Fig. 4-9. By selecting the proper value of R_2, we can make the resistance seen by the base bias current I_{B_2} equal to the resistance seen by the base bias current I_{B_1}. This will raise input 2 to the dc voltage at input 1. In other words, if the currents I_{B_1} and I_{B_2} are equal, and the resistances seen by these currents are equal, the voltages to ground at inputs 1 and 2 are equal. This means that there will be no dc differential input voltage V_{id} to cause an output offset V_{oo}. The value of R_2 needed to eliminate or reduce the dc differential input voltage, V_{id}, and the resulting output offset, V_{oo}, is easily found.

Note in Fig. 4-9 that current I_{B_1} sees two parallel paths, one containing R_1 and R_s in series and the other containing R_F. Thus when $V_o \cong 0\,\text{V}$, current I_{B_1} sees a resistance $R_1 + R_s$ in parallel with the feedback resistor R_F. Since current I_{B_2} is to see a resistance R_2 that is equal to the total resistance seen by I_{B_1}, we can show that

$$R_2 = \frac{(R_1 + R_s)R_F}{R_1 + R_s + R_F} \tag{4-5}$$

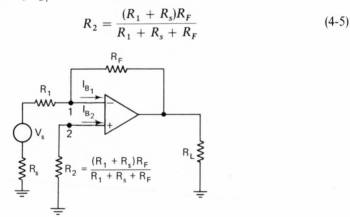

Fig. 4-9 Resistor R_2 in series with the noninverting input reduces the output offset voltage caused by the input bias current I_B.

EXAMPLE 4-4

Referring to the circuit in Fig. 4-6, what value of resistance can we use in series with the noninverting input (pin 3) to reduce or eliminate the output offset voltage caused by the input bias current? Assume that the internal resistance of the signal source V_s is negligible.

Answer. Since the input resistor $R_1 = 1\,\text{k}\Omega$, the feedback resistor $R_F = 100\,\text{k}\Omega$, and the signal source's resistance $R_s = 0\,\Omega$, we can use

$$R_2 = \frac{(1\,\text{k}\Omega + 0)100\,\text{k}\Omega}{1\,\text{k}\Omega + 0 + 100\,\text{k}\Omega} = 990\,\Omega$$

4-3   Input Offset Current I_{io}

In the previous section we learned that a resistor R_2 placed in series with
the noninverting input 2 reduces the output offset caused by the input bias
current. However, Eq. (4-5) for finding the necessary value of R_2 was derived
assuming that the base bias currents I_{B_1} and I_{B_2} are equal. In practice, due
to imbalances within the Op Amp's circuitry, these currents are at best only
approximately equal, as indicated in Eq. (4-3). The input offset current I_{io},
usually specified by the manufacturer, is a parameter that indicates how far
from being equal the currents I_{B_1} and I_{B_2} can be. In fact, the input offset
current is defined as the difference in the two base bias currents, that is,

$$I_{io} = |I_{B_1} - I_{B_2}| \qquad (4\text{-}6)$$

When given the value of input offset current I_{io}, we can predict how much
output offset voltage a circuit like that in Fig. 4-9 might have, caused by the
existence of base bias currents. Due to inequalities of the currents I_{B_1} and
I_{B_2}, the voltages with respect to ground at inputs 1 and 2 will be unequal,
even though a properly chosen value of R_2 is used. This causes a dc differential input voltage V_{id}, which in turn causes an output offset voltage V_{oo}.
In other words, the amount of dc differential input voltage and the amount
of resulting output offset depend on the amount of difference in the base bias
currents I_{B_1} and I_{B_2}, which is the input offset current I_{io}. More specifically,
the amount of output offset voltage V_{oo} in a circuit like that in Fig. 4-9,
caused by the input offset current I_{io}, can be closely approximated with the
equation*

$$V_{oo} \cong R_F I_{io} \qquad (4\text{-}7)$$

if the value of R_2 was determined with Eq. (4-5).

EXAMPLE 4-5

Referring to the circuit in Fig. 4-10, find the maximum output offset
caused by the input offset current I_{io}. The Op Amp is a type 741, and
the internal resistance of the signal source V_s is negligible.

Answer. The 741's maximum input offset current $I_{io} = 200\,\text{nA}$ as
indicated on its specification sheet. Since the value of R_2 satisfies

---

* See Appendix A2 for the derivation.

Eq. (4-5), we can closely estimate the output offset caused by I_{io} with Eq. (4-7). In this case

$$V_{oo} \cong R_F I_{io} = 1 \text{ M}\Omega(200 \text{ nA}) = 200 \text{ mV}$$

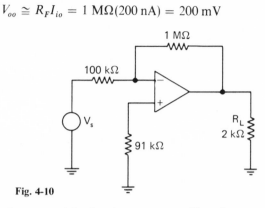

Fig. 4-10

This shows that we might have an output offset due to unequal base bias currents even though a properly chosen resistor R_2 is placed between the noninverting input and ground. If R_2 is not used, however, the output offset caused by input currents is considerably larger and therefore potentially a greater problem.

EXAMPLE 4-6

Referring again to the circuit in Fig. 4-10, find the maximum output offset caused by the input currents if R_2 is removed and, instead, the noninverting input 2 is connected directly to ground. As before, the Op Amp is a type 741.

Answer. $V_{oo} \cong R_F I_B = 1 \text{ M}\Omega(500 \text{ nA}) = 500 \text{ mV}$.

It is interesting to note that the output offset can be more than doubled if the resistor R_2 is not used. Therefore, R_2 may or may not be necessary, depending on how important it is to have little or no output offset. The required stability of the output voltage with temperature changes must be considered too, as we will see in a later chapter. The size of the feedback resistor is also a factor. As mentioned before, if a relatively small feedback resistor is used, the effects of base bias current and the input offset current are also relatively small. Some types of IC Op Amps are made with FETs or high-beta transistors* in the input differential stage. These have dc input

---

* Op Amps with high-beta input transistors are sometimes called *super beta* transistor input Op Amps. An extremely high beta can be achieved by connecting conventional transistors in a Darlington pair.

currents on the order of just a few nonoamperes. Some *hybrid* models of Op Amps, which contain discrete and integrated circuits, have input bias currents as low as 0.01 pA. Of course, such a small dc input bias current reduces the current-generated output offset voltage to insignificance. As we will see in applications later, an extremely small input bias current is desirable, and in fact necessary, in long-term integrating and sample-and-hold circuits.

4-4   Combined Effects of V_{io} and I_{io}

In a circuit such as that in Fig. 4-9, the input offset voltage V_{io} can cause either a positive or a negative output offset voltage V_{oo}. See Eq. (4-1). Likewise, the input offset current I_{io} can cause either a positive or a negative output offset voltage V_{oo}. See Eq. (4-7). The effects of input offset voltage V_{io} and input offset current I_{io} might buck and cancel each other, resulting in little output offset. On the other hand, they might be additive, causing an output offset voltage V_{oo} that is the sum of the output offsets caused by V_{io} and I_{io} working independently. Thus the total output offset voltage in the circuit of Fig. 4-9 can be as large as

$$V_{oo} \cong A_v V_{io} + R_F I_{io} \qquad (4\text{-}8a)$$

if the value of R_2 satisfies Eq. (4-5). This equation is sometimes shown in other equivalent forms, such as

$$V_{oo} \cong [V_{io} + R_2 I_{io}]\left(\frac{R_F}{R_1} + 1\right) \qquad (4\text{-}8b)$$

or

$$V_{oo} \cong [V_{io} + R_2(I_{B_1} - I_{B_2})]\left(\frac{R_F}{R_1} + 1\right) \qquad (4\text{-}8c)$$

EXAMPLE 4-7

Considering the effects of both V_{io} and I_{io}, what is the maximum output offset voltage V_{oo} of the circuit in Fig. 4-10 if the Op Amp is a type 741?

Answer. Arbitrarily selecting Eq. (4-8b) above, we can show that the maximum output offset voltage is

$$V_{oo} \cong [6\,\text{mV} + 91\,\text{k}\Omega(200\,\text{nA})]\left(\frac{1000 + 100}{100}\right) = 266\,\text{mV}$$

REVIEW QUESTIONS

1. What is the input offset voltage of an Op Amp?

2. Why is an input offset voltage V_{io} often needed to null the output of an Op Amp?

3. What polarity of required input offset voltage would you expect at input 1 with respect to input 2?

4. What is the meaning of the term *base bias current*?

5. What are the difference and similarity of the base bias and input bias currents?

6. What problem might exist if a significant base bias current exists and if the feedback resistor is relatively large?

7. How can the undesirable effect of a significant input bias current be minimized?

8. The nulling circuits in Figs. 4-2 and 4-3 are able to provide a dc differential input voltage that is (*positive*), (*negative*), (*either polarity*) at input 1 with respect to input 2.

9. The nulling circuits in Figs. 4-2 and 4-3 are able to adjust the output V_o to zero volts in spite of the existence of (*input offset voltage*), (*input bias current*), (*input offset voltage and input bias current*).

10. What is the meaning of the term *input offset current*?

PROBLEMS

1. If the Op Amp in the circuit of Fig. 4-11 is a type 741, what is the circuit's closed-loop voltage gain if the potentiometer (POT) is adjusted to 39 kΩ? The 8-pin DIP (dual-in-line) package has the same pin identifications as does the 8-pin circular metal can package.

2. If the Op Amp in the circuit of Fig. 4-11 is a type 777 (Appendix SIV), what is the circuit's closed-loop gain if the POT is adjusted to 14 kΩ? The 8-pin DIP (dual-in-line) package has the same pin identifications as does the 8-pin circular metal can package.

3. If the slide on the POT is moved to the far right in the circuit of Fig. 4-11, what is the approximate closed-loop voltage gain of the circuit?

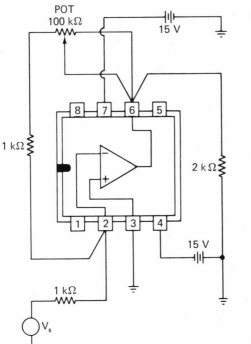

Fig. 4-11

4. If the slide on the POT is moved to the far left in the circuit of Fig. 4-11, what is the approximate closed-loop voltage gain of the circuit?

5. In the circuit described in Problem 1, what is the maximum output offset caused by the Op Amp's input offset voltage?

6. What is the maximum output offset caused by the input offset voltage in the circuit described in Problem 2?

7. In Problem 1, what is the circuit's maximum output offset caused by the existence of base bias currents in the input leads?

8. What is the maximum possible output offset voltage caused by the existence of input bias currents in the circuit described in Problem 2?

9. Refer to the circuit described in Problem 1. What maximum output offset voltage might it have with the combined effects of input offset voltage and input bias currents?

10. Refer to the circuit described in Problem 2. What maximum output offset voltage might it have with the combined effects of input offset voltage and input bias currents?

11. If the POT in the circuit of Fig. 4-11 is replaced with a fixed 100-kΩ resistor, what value of resistance should we put in series with pin 3 and ground to reduce the effect of input bias currents? Assume that the Op Amp is a type 741 and that the internal resistance of the signal source is negligible.

12. If the POT in the circuit of Fig. 4-11 is replaced with a fixed 10-kΩ resistor, what value of resistance should we put in series with pin 3 and ground to reduce the effect of input bias current? The Op Amp is a type 777 and the internal resistance of the signal source is negligible.

13. In the circuit described in Problem 11, what is the maximum output offset voltage caused by the existence of input bias currents if the properly selected resistance is used between pin 3 and ground?

14. In the circuit described in Problem 12, what is the maximum output offset caused by the existence of input bias currents if the properly selected resistance is used between pin 3 and ground?

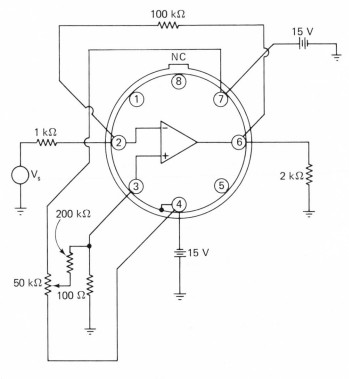

Fig. 4-12

15.   In the circuit of Fig. 4-12, if adjustment of the 50 kΩ POT almost, but not quite, nulls the output, what small modification can we make to have full null control?

16.   If in the circuit of Fig. 4-12 the input signal V_s is a sine wave but the output has the waveform shown in Fig. 4-13, what modification or adjustment will probably correct the problem?

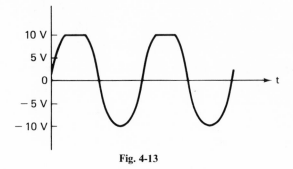

Fig. 4-13

17.   If the circuit in Fig. 4-12 has a maximum peak-to-peak output capability of 20 V, what maximum peak-to-peak input signal V_s can we apply?

18.   If the circuit in Fig. 4-12 is nulled, has a maximum peak-to-peak output capability of 20 V, and has an input sine-wave voltage V_s of 2 V peak-to-peak, what modification is necessary in this circuit if clipping of the output signal is to be avoided?

19.   If V_s is a sine wave in the circuit of Fig. 4-14, and if an oscilloscope reading shows a sine-wave voltage at pin 6 to ground with a peak-to-peak value of 6.3 V, what is the peak-to-peak amplitude of the input V_s?

20.   If the circuit in Fig. 4-14 has the waveform shown in Fig. 4-13 at pin 6 with respect to ground, what is the approximate peak-to-peak amplitude of V_s?

21.   What is the purpose of the 10-kΩ POT in the circuit of Fig. 4-14 if the Op Amp is a type 741?

22.   If the Op Amp in the circuit of Fig. 4-14 is a type 741, what is its maximum peak-to-peak output signal capability?

23.   If the Op Amp in the circuit of Fig. 4-15 is a type 741, what is its maximum possible output offset, caused by both the input offset voltage and the input offset current?

24. If the Op Amp in the circuit of Fig. 4-15 is a type 777, what is its maximum possible output offset, caused by both the input offset voltage and the input offset current?

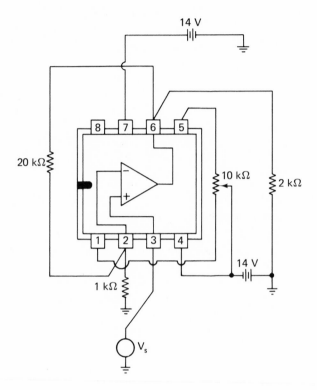

Fig. 4-14

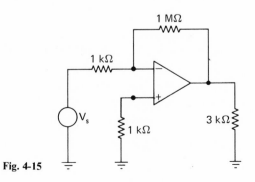

Fig. 4-15

CHAPTER 5

COMMON-MODE
VOLTAGES
AND DIFFERENTIAL
AMPLIFIERS

A voltage appearing at both inputs of an Op Amp, as shown in Fig. 5-1, is a common-mode voltage V_{cm}. A common-mode voltage V_{cm} can be dc, ac, or a combination (superposition) of dc and ac. Thus a common-mode voltage V_{cm} might be an unavoidable dc level on both inputs.

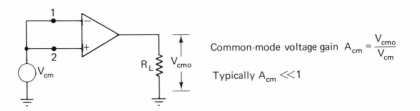

Common-mode voltage gain $A_{cm} = \dfrac{V_{cmo}}{V_{cm}}$

Typically $A_{cm} \ll 1$

Fig. 5-1 Op Amp with common-mode voltage applied.

When Op Amps are required to work in time-varying magnetic and electric fields, ac voltage can be induced into both input leads. In such cases, dc and ac common-mode voltages V_{cm} can exist simultaneously at an Op Amp's inputs. The induced ac voltage is usually 60 Hz from nearby electrical power equipment, but it can be higher frequencies or even irregular transients. In any case, it is undesired and considered as *noise* in the system.

Amplifiers with differential inputs (e.g., Op Amps) have more or less ability to reject common-mode voltages. This means that, although fairly large common-mode voltages might exist at an Op Amp's differential inputs, these voltages can be reduced to very small and often insignificant amplitudes at the output. In this chapter we will become familiar with manufacturers' specifications that will enable us to predict how well a given Op Amp will reject (not pass) common-mode voltages. We will also see applications in which the Op Amp's ability to reject common-mode voltages is especially useful.

5-1 The Differential-Mode Op Amp Circuit

To fully appreciate how a properly wired Op Amp circuit is able to reject undesirable common-mode voltages, we should first look at the familiar circuits in which induced noise is potentially a problem. For example, both circuits shown in Fig. 5-2, voltage V_s is the desired signal to be amplified and is typically from a preceding transducer or amplifier. If the input lead has

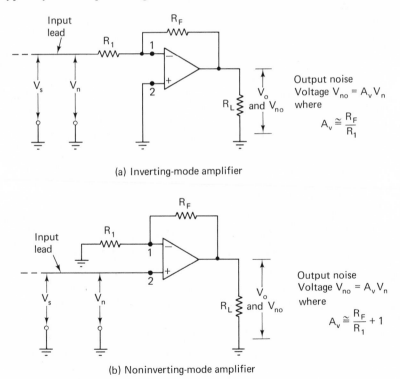

(a) Inverting-mode amplifier

Output noise Voltage $V_{no} = A_v V_n$ where

$$A_v \cong \frac{R_F}{R_1}$$

(b) Noninverting-mode amplifier

Output noise Voltage $V_{no} = A_v V_n$ where

$$A_v \cong \frac{R_F}{R_1} + 1$$

Fig. 5-2 Inverting-mode and noninverting-mode circuits amplify induced noise.

any appreciable length, and if varying fields are present, noise voltage V_n will probably be induced into it as shown. In both of these circuits, the Op Amp cannot distinguish noise V_n from desired signal V_s; both are amplified and appear at the output. Thus if the closed-loop gain A_v of each of these circuits is about 100, the output noise voltage V_{no} and output signal voltage V_o are about 100 times larger than their respective inputs.

Noise voltages at the output of an Op Amp are greatly reduced if the circuit is connected to operate in a differential mode, as shown in Fig. 5-3a. In this circuit, the desired signal V_s is amplified normally because it is applied *across* the two inputs. That is, signal V_s causes a differential input V_{id} to appear across the input terminals 1 and 2. The noise voltage V_n, however, is induced into each input lead with respect to ground or a common point.

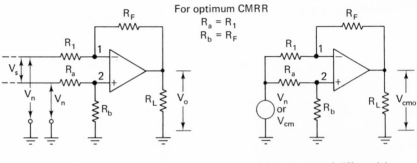

(a) Op Amp connected in differential mode; $A_V \cong -\dfrac{R_F}{R_1}$.

(b) Equivalent of differential-mode circuit showing induced noise voltage V_n as a common-mode voltage V_{cm}.

Fig. 5-3

With proper component selection, both of the induced noise voltages V_n are equal in amplitude and phase and therefore are common-mode voltages as shown in Fig. 5-3b. Thus if the noise voltage to ground at input 1 is the same as the noise voltage to ground at input 2, the differential noise voltage across input terminals 1 and 2 is zero. Since ideally an Op Amp amplifies only differential input voltages, no noise voltage should appear at the output. However, due to imperfections within the practical Op Amp, such as the fact that the current source (Q_3 in Fig. 1-8) of the first differential stage is not a perfect constant-current source, both emitter and collector currents of this stage increase or decrease when a more positive or a more negative voltage is applied to both inputs 1 and 2 simultaneously. Thus, in practice, some output common-mode voltage V_{cm} will get through the Op Amp to the load R_L. The ratio of the output common-mode voltage V_{cmo} to the input common-mode voltage V_{cm} is called the common-mode gain A_{cm}, that is,

$$A_{cm} = \frac{V_{cmo}}{V_{cm}} \qquad (5\text{-}1)$$

Ideally, this gain A_{cm} is zero. In practice, it is finite but much smaller than 1.

5-2 Common-Mode Rejection Ratio, CMRR

Op Amp manufacturers usually do not list a common-mode gain factor. Instead, they list a *common-mode rejection ratio, CMRR*. The *CMRR* is defined in several essentially equivalent ways by the various manufacturers. It can be defined as the ratio of the change in the input common-mode voltage V_{cm} to the resulting change in input offset voltage V_{io}. Thus

$$CMRR = \frac{\Delta V_{cm}}{\Delta V_{io}} \qquad (5\text{-}2)$$

It can also be shown to be approximately equal to the ratio of the closed-loop gain A_v to the common-mode gain A_{cm}, that is,

$$CMRR = \frac{A_v}{A_{cm}} \qquad (5\text{-}3)$$

Generally, larger values of *CMRR* mean better rejection of common-mode signals and are therefore more desirable in applications where induced noise is a problem and the differential-mode circuit is used.

The common-mode rejection is usually specified in decibels (dB), where

$$CMR(\text{dB}) = 20 \log \frac{\Delta V_{cm}}{\Delta V_{io}} \qquad (5\text{-}4a)$$

or

$$CMR(\text{dB}) = 20 \log \frac{A_v}{A_{cm}} = 20 \log CMRR \qquad (5\text{-}4b)$$

A chart for converting common-mode rejection from a ratio to dBs, or vice versa, is given in Fig. 5-4.

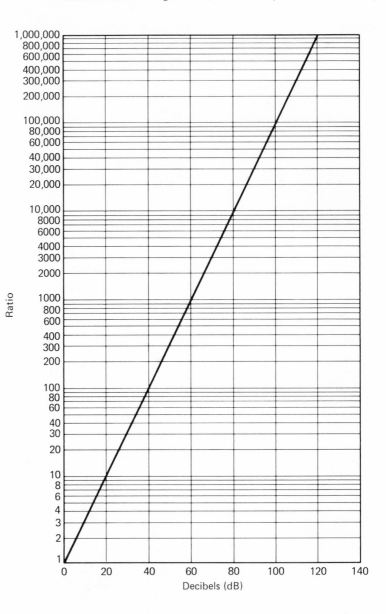

Fig. 5-4 Chart for converting voltage ratios to dB and vice versa.

Equation (5-3) shows that the common-mode voltage gain A_{cm} of a differential-mode Op Amp circuit is a function of its closed-loop gain and the Op Amp's specified *CMRR*. Rearranging Eq. (5-3), we can show that

$$A_{cm} = \frac{A_v}{CMRR} \tag{5-3}$$

The equivalent of this in dBs can be shown as

$$20 \log A_{cm} = 20 \log A_v - 20 \log CMRR \tag{5-5a}$$

or as

$$A_{cm}(\text{dB}) = A_v(\text{dB}) - CMR(\text{dB}) \tag{5-5b}$$

Since the closed-loop gain A_v is usually much smaller than the *CMRR*, the common-mode gain A_{cm} in Eq. (5-3) is much smaller than 1. In either version of Eq. (5-5) then, the common-mode gain in dBs is negative.

Some manufacturers define *CMRR* as the reciprocal of the right side of Eq. (5-2). This implies that the *CMRR* in dB is a negative value. If we assume that the specified value of *CMR*(dB) is positive, then we subtract it from A_v(dB) to obtain the value of A_{cm}(dB), as shown in Eq. (5-5b). If we assume that *CMR*(dB) is negative, then we add it to A_v(dB) to get A_{cm}(dB). In either case, we solve for the *difference* in the values on the right side of Eq. (5-5) to determine A_{cm}(dB).

The signal voltage gain V_o/V_s of the differential circuit in Fig. 5-3a is determined with Eq. (3-4), that is,

$$A_v = \frac{V_o}{V_s} \cong -\frac{R_F}{R_1} \tag{3-4}$$

As shown in Fig. 5-3a, the resistors R_a and R_b are selected to be equal to R_1 and R_F, respectively. This assures us of equal resistances to ground, looking into either input lead. It also provides equal induced noise voltages to ground at input terminals 1 and 2 and thus makes noise voltages common mode, enabling the Op Amp to reject the induced noise, more or less depending on its *CMRR* value. Resistors R_a and R_b also serve the same function as R_2 does in the circuit of Fig. 4-9, i.e., to reduce the output offset V_{oo} caused by input bias current I_B.

EXAMPLE 5-1

Referring to the circuit in Fig. 5-3a, if $R_1 = R_a = 1$ kΩ, $R_F = R_b = 10$ kΩ, $V_s = 10$ mV at 1000 Hz, and $V_n = 10$ mV at 60 Hz, what are the amplitudes of the 1000-Hz signal and the 60-Hz noise at the load R_L? The Op Amp's *CMR*(dB) = 80 dB.

Answer. This circuit's closed-loop gain is

$$A_v \cong -\frac{R_F}{R_1} = -\frac{10 \text{ k}\Omega}{1 \text{ k}\Omega} = -10$$

The negative sign means that the signal voltage output with respect to ground is out of phase with the signal at the inverting input 1 with respect to the noninverting input 2. Since the 1000 Hz is applied in a differential mode, it is amplified by this gain factor. Thus at 1000 Hz, the output signal is

$$V_o \cong -10(10 \text{ mV}) = -100 \text{ mV}$$

The 60-Hz noise voltage, on the other hand, is applied in common mode. Therefore, we can first find the common-mode gain with Eq. (5-3), that is,

$$A_{cm} = \frac{A_v}{CMRR} \cong \frac{-10}{10,000} = -1 \times 10^{-3}$$

where 80 dB is equal to 10,000 according to Fig. 5-4.

Since the 60-Hz noise is the common-mode voltage, we can find the amount of output common-mode voltage with Eq. (5-1), that is,

$$V_{cmo} = A_{cm}V_{cm} \cong -1 \times 10^{-3}(10 \text{ mV}) = -10 \text{ } \mu\text{V}$$

Thus, the 60-Hz output is smaller than the input induced 60 Hz, by the factor 1000, and the Op Amp is shown as an effective tool in reducing noise problems.

We could have determined the common-mode output V_{cmo} using Eqs. (5-5). For example, since a closed-loop gain of 10 is equivalent to 20 dB according to Fig. 5-4, then by Eq. (5-5b),

$$A_{cm}(\text{dB}) = 20 - 80 = -60 \text{ dB}$$

The negative sign means that the common-mode voltage is attenuated by 60 dB, which means that V_{cmo} is smaller than V_{cm} by the factor 1000. Note that 60 dB is equivalent to the ratio 1000 in Fig. 5-4. Thus $V_{cmo} = V_{cm}/1000$, etc.

5-3   Maximum Common-Mode Input Voltages

As we would expect, there are limits on how much positive or negative voltage can be applied to both inputs of a practical Op Amp. Certainly, excessive input voltages with respect to the ground or a common point of the Op Amp can ruin the input differential stage or at least drive it to saturation and cause distortion of the output signals. If we expect relatively large amplitudes of common-mode input voltages, which might be either ac or dc levels, we must use relatively large dc supply voltages.

In the circuit of Fig. 5-5, the maximum voltages at inputs 1 and 2 are not as large as the maximum voltages applied to inputs I and II. The voltage V_2 is a fraction of the applied voltage V_{II} by voltage-divider action of R_a and R_b. Therefore, assuming that the Op Amp is ideal,

$$V_2 = \frac{V_{II}R_b}{R_a + R_b} \tag{5-6}$$

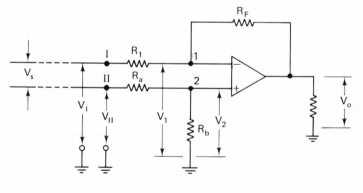

Fig. 5-5

Since V_1 is *virtually* at the same potential as V_2, then we can similarly show that

$$V_1 \cong \frac{V_{II}R_b}{R_a + R_b} \tag{5-7}$$

The voltage V_1 or V_2 should never exceed the maximum *input voltage* or the maximum *common-mode voltage* specified on an Op Amp's data sheets. Typical maximum input (common-mode) voltage vs dc supply voltage characteristics are shown in Fig. 5-6. If V_I and V_{II} in Fig. 5-5 are dc levels or in-phase ac voltages, the common-mode voltage at inputs I and II is the average of these voltages, that is,

$$V_{cm} = \frac{V_I + V_{II}}{2} \tag{5-8}$$

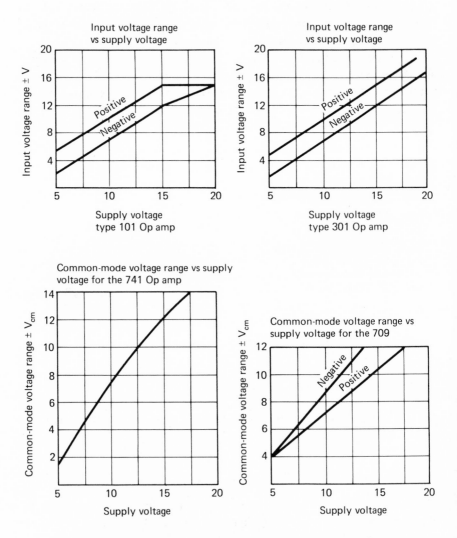

Fig. 5-6 Typical input (common-mode) voltages vs dc supply voltages.

5-4   Op Amp Instrumentation Circuits

Frequently, in instrumentation and industrial applications, the Op Amp is used to amplify signal output voltages from bridge circuits. We considered such a circuit earlier in Fig. 3-17a. Similar circuits are the useful *half-bridge* circuits as shown in Fig. 5-7. Because these circuits are single-ended, i.e., each of their Op Amps is connected to work in an inverting mode or a noninverting mode, they are prone to induced noise problems. Therefore, if the bridge-type transducer circuit is required to work in time-varying

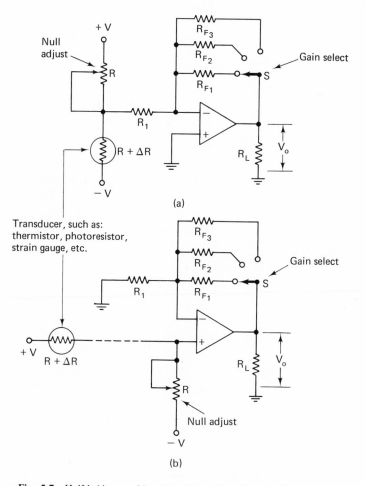

(a)

(b)

Fig. 5-7 Half-bridge working into (a) an inverting circuit and (b) a noninverting circuit.

electric and magnetic fields, its associated amplifier should be a differential type, such as in Fig. 5-8.

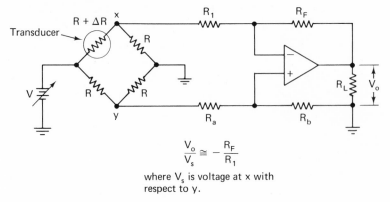

$$\frac{V_o}{V_s} \cong -\frac{R_F}{R_1}$$

where V_s is voltage at x with respect to y.

$$R_1 = R_a$$
$$R_F = R_b$$
for good CMRR

Fig. 5-8 Bridge working into a differential-mode amplifier has good *CMRR*.

In the circuit of Fig. 5-8, each of the points x and y has a dc potential to ground which is some fraction of the bridge's dc source voltage V. The sensitivity of the bridge increases with larger values of V, but ther large values of V cause larger dc common-mode voltages. The average of the dc voltages at points x and y is the dc common-mode input voltage to the Op Amp. Therefore the maximum recommended input common-mode voltage dictates the maximum allowable dc source voltage V on the bridge. The transducer's resistance changes by ΔR when the appropriate change in its physical environment occurs. The bridge converts this physical change to an electrical voltage change across points x and y which is amplified by the Op Amp. Thus, depending on the type of transducer used, a change in temperature, light intensity, strain, etc., is converted to an amplified electrical signal. The gain of the differential amplifier can be increased or decreased if we increase or decrease *both* R_F and R_b by equal amounts. In other words, R_F and R_b must remain equal if good *CMR* is to be retained. However, if R_F and R_b are changed, the resistances looking into points x and y will change noticeably too. The accompanying problems with this kind of variable loading on the bridge can be avoided by use of high input impedance circuits such as in Fig. 5-9. The gain factors of these circuits are shown to be negative, representing the out-of-phase relationship of V_o and V_s, where V_s is measured from input I with respect to input II.

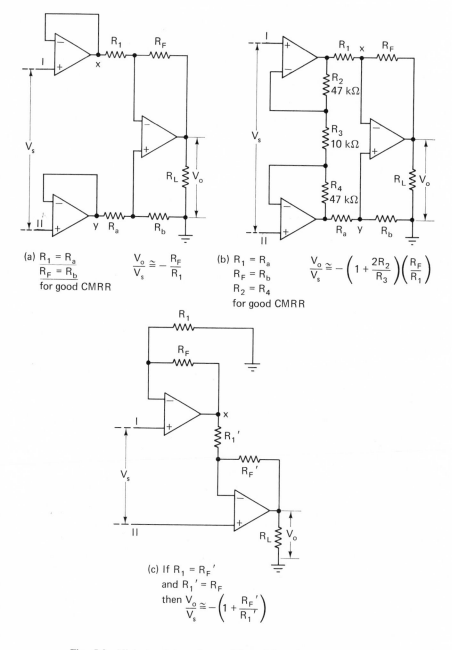

(a) $R_1 = R_a$
$R_F = R_b$
for good CMRR

$$\frac{V_o}{V_s} \cong -\frac{R_F}{R_1}$$

(b) $R_1 = R_a$
$R_F = R_b$
$R_2 = R_4$
for good CMRR

$$\frac{V_o}{V_s} \cong -\left(1 + \frac{2R_2}{R_3}\right)\left(\frac{R_F}{R_1}\right)$$

(c) If $R_1 = R_F'$
and $R_1' = R_F$
then $\frac{V_o}{V_s} \cong -\left(1 + \frac{R_F'}{R_1'}\right)$

Fig. 5-9 High input impedance differential-mode amplifiers. The negative signs in the gain equations mean phase reversal if V_s is measured from the upper input with respect to the lower input.

The circuit shown in Fig. 5-9a is simply a differential-mode amplifier preceded by a voltage follower at each input. These voltage followers have extremely large input impedances (resistances) and therefore draw negligible current from the source of signal V_s. Thus, even if the gain of the differential stage is changed, the resistance into inputs I and II remains large and constant. Since voltage followers have unity gain, the gain of the entire circuit is the gain of the output differential stage. Thus, for the circuit shown in Fig. 5-9a,

$$\frac{V_o}{V_s} \cong -\frac{R_F}{R_1} \tag{3-4}$$

Some manufacturers of Op Amps recommend the circuit in Fig. 5-9b as a high input impedance differential amplifier. It is a modified version of the circuit in Fig. 5-9a. The circuit in b of the figure differs in that the Op Amps at the inputs are connected not as voltage followers, but instead as non-inverting amplifiers. Instead of having 100% feedback as voltage followers do, the input noninverting amplifiers of the circuit in Fig. 5-9b have partial feedback, and therefore each has a voltage gain greater than unity. The combined gain of both input noninverting stages can be determined with the expression $(1 + (2R_2)/R_3)$. Thus the voltage across x and y is larger than the differential input voltage across inputs I and II by this gain factor. The voltage across x and y is then further amplified by the output differential stage, which is by the factor $-(R_F/R_i)$. The total gain of the circuit in Fig. 5-9b is the product of the stage gains. Thus

$$\frac{V_o}{V_s} \cong -\left(1 + \frac{2R_2}{R_3}\right)\left(\frac{R_F}{R_1}\right) \tag{5-9}$$

This circuit's gain can be varied, while retaining good $CMRR$ capability, by use of a potentiometer for R_3. Larger or smaller values of R_3 will give smaller or larger total circuit gains, respectively.

The circuit in Fig. 5-9c also has high input impedance due to a non-inverting-mode amplifier at each input. This circuit's differential gain V_o/V_s is determined by the components connected on the output (lower) Op Amp, namely, R_1' and R_F'. Once the gain and components R_1' and R_F' are selected, the components in the upper stage must comply with the equations

$$R_F = R_1'$$

and

$$R_1 = R_F'$$

for good *CMRR*. For example, suppose that we select $R'_1 = 1\,\text{k}\Omega$ and $R'_F = 100\,\text{k}\Omega$ for a gain of 101 according to Eq. (3-6). Then the upper stage, $R_1 = 100\,\text{k}\Omega$ and $R_F = 1\,\text{k}\Omega$; the gain, by the same equation, is 1.01. Suppose now that a $+5$-V dc common-mode voltage is applied to inputs I and II. The lower stage *tends* to amplify the $+5$ V at input II by 101. The upper stage does amplify the $+5$ V at input I by 1.01, resulting in an output at point x that is $+5.05$ V. However, the output of the upper stage sees the lower stage as a noninverting type whose gain is $-100\,\text{k}\Omega/1\,\text{k}\Omega = -100$. Thus the lower stage *tends* to amplify the $+5.05$ V at point x by -100, while it simultaneously *tends* to amplify the $+5$ V at point II by 100. The final output voltage V_o is therefore the superposition of these two effects, that is, the output V_o is the sum of the outputs caused by the inputs at points x and II. Thus

$$V_0 = 101\,(5\text{ V}) + (-100)\,5.05\text{ V} = 0\text{ V}$$

which illustrates the *CMRR* capabilities of the circuit in Fig. 5-9c. Its differential gain is

$$\frac{V_o}{V_s} \cong -\left(1 + \frac{R'_F}{R'_1}\right) \tag{5-10}$$

Another variable-gain differential-mode amplifier is shown in Fig. 5-10. Its gain V_o/V_s is varied by an adjustment of the potentiometer R_3, and its *CMRR* is not appreciably affected thereby. Generally, the gain is increased or decreased by a decrease or increase, respectively, of R_3. In this circuit, the right end of the feedback resistor R_F is terminated with a voltage divider and

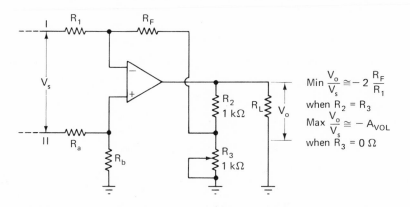

Fig. 5-10 Differential amplifier with variable gain.

the output signal voltage V_o is across this divider. When R_3 is maximum, 1 kΩ in this case, half as much signal is fed back to the inverting input via R_F than would be the case if R_F were connected directly to the output as in the simpler differential amplifier in Fig. 5-3a. Thus the minimum gain can be estimated with the equation:

$$\frac{V_o}{V_s} \cong -2\left(\frac{R_F}{R_1}\right) \qquad (5\text{-}11)$$

This shows that twice as much gain occurs with half as much negative feedback. If R_3 is adjusted to a minimum of 0 Ω, the right end of R_F is grounded and no negative feedback occurs. This tends to pull the stage gain up to the open-loop gain A_{VOL} of the Op Amp. Of course, both inputs can be preceded with voltage followers if high input impedance is required.

REVIEW QUESTIONS

1. What is a common-mode voltage?

2. What is the ideal value of $CMRR$?

3. If an Op Amp is to work in an environment that has significant induced voltages, does grounding *one* of the inputs reduce noise in the output? Why?

4. What do the initials $CMRR$ mean?

5. Graphically, convert the following $CMRR$ values to decibels: 10, 100, 1000, 10,000, 100,000, and 10^6.

6. What is the ideal value of common-mode voltage gain?

7. In what configuration is the Op Amp arranged to make induced noise voltages common mode?

8. What effect do the dc supply voltage values have on the maximum recommended common-mode voltages?

9. Assume that in the circuit of Fig. 5-3a, R_F is replaced with a potentiometer, while all other resistors remain fixed. What will happen to the circuit's differential gain A_v and its $CMRR$ capability if the potentiometer is varied?

10. What is the purpose of using a voltage follower in each input of a differential amplifier?

PROBLEMS

1. In the circuit of Fig. 5-11, the output V_{cmo} measures 4 mV rms, 60 Hz, when the applied $V_{cm} = 400$ mV rms, 60 Hz. What is the Op Amp's common-mode rejection in decibels?

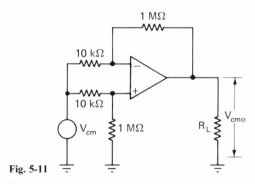

Fig. 5-11

2. In the circuit of Fig. 5-11, the output V_{cmo} measures about 127 μ V rms, 60 Hz, when the applied $V_{cm} = 400$ mV rms, 60 Hz. What is this Op Amp's common-mode rejection in decibels?

3. In the circuit of Fig. 5-12, if the signal voltage at input I varies from -30 mV up to 20 mV and if 10 mV rms, 60 Hz, is present at input I to ground, what are (a) the signal variations at output 0 and (b) the 60-Hz rms voltage V_n at output 0? The Op Amp's $CMRR = 10,000$.

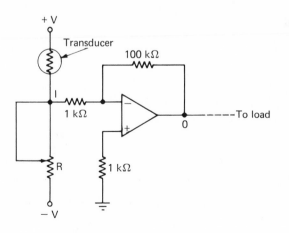

Fig. 5-12

4. If the signal at input I to ground in the circuit of Fig. 5-12 varies from $-25\,\text{mV}$ to $+35\,\text{mV}$ while $15\,\text{mV}$ rms, $60\,\text{Hz}$ is also present at this input, what are (a) the signal variations and (b) the 60-Hz rms voltage V_n at output 0? The Op Amp's $CMRR = 15,000$.

5. Refer to the circuit in Fig. 5-13. If $R_F = R_b = 10\,\text{k}\Omega$, $R_1 = R_a = 1\,\text{k}\Omega$, $CMR(\text{dB}) = 80\,\text{dB}$, signal across inputs I and II varies between $-0.5\,\text{V}$ and $+0.5\,\text{V}$, and if both of these inputs have $+2\,\text{V}$ dc and $20\,\text{mV}$ rms, $60\,\text{Hz}$, potentials with respect to ground, find (a) the approximate dc voltages to ground at inputs 1 and 2, (b) the output signal variations, and (c) the rms value of 60-Hz voltage at output 0.

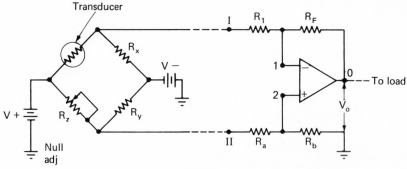

Fig. 5-13

6. If in the circuit of Fig. 5-13 $R_F = R_b = 20\,\text{k}\Omega$, $R_1 = R_a = 500\,\Omega$, $CMRR = 50,000$, signal across inputs I and II varies from $-80\,\text{mV}$ to $+80\,\text{mV}$, and if both of these inputs have $+4\,\text{V}$ dc and $600\,\text{mV}$ rms, $60\,\text{Hz}$, potentials to ground, find (a) the approximate dc voltages to ground at inputs 1 and 2, (b) the output signal variations, and (c) the rms value of the 60-Hz voltage at output 0.

7. If the Op Amp in the circuit of Fig. 5-13 is a type 301, and if its dc supply voltages are $\pm 15\,\text{V}$, to what maximum positive voltage and to what maximum negative voltage can we drive both inputs 1 and 2? Refer to Fig. 5-6.

8. To what maximum positive or negative voltage can we drive both inputs of the 741 Op Amp if its dc supply voltages are $\pm 17.5\,\text{V}$?

9. If $R_y = R_z$ in the circuit of Fig. 5-14, and if the Op Amps are 741s with dc supply voltages of $\pm 15\,\text{V}$, to what maximum value can we adjust voltage V and still not exceed the maximum recommended input voltage of any of these Op Amps?

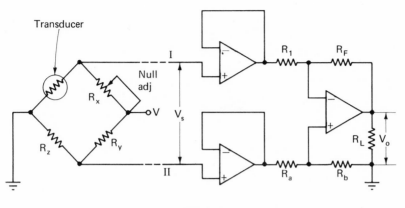

Fig. 5-14

10. If $R_y = R_z$ in the circuit of Fig. 5-14, and if the Op Amps are 101 types with dc supply voltages of ± 15 V, to what maximum positive value can we adjust voltage V and still not exceed the maximum recommended input voltage of any of these Op Amps?

11. If voltage $V = +5$ V to ground and $R_y = R_z$ in the circuit of Fig. 5-14, what are the maximum positive and negative values of V_s if the transducer's resistance can vary from zero to infinity? Assume $R_x > 0$.

12. If voltage $V = +6$ V to ground and if $R_y = 2R_z$ in the circuit of Fig. 5-14, what are the maximum positive and negative values of V_s if the transducer's resistance can vary from zero to infinity? Assume $R_x > 0$.

13. In Problem 11, if V_o clips at ± 12 V, what maximum ratios of R_F/R_1 R_b/R_a can we use and still avoid such clipping?

14. In Problem 12, if V_o clips at ± 12 V, what maximum ratios of R_F/R_1 and R_b/R_a can we use and still avoid such clipping?

15. If in the circuit of Fig. 5-14, $R_F = R_b = 100$ kΩ, $R_1 = R_a = 2$ kΩ, and the output signal V_o clips if we attempt to drive it beyond ± 10 V, what maximum V_s can we apply and still avoid such clipping?

16. In the circuit of Fig. 5-14, if $R_F = R_b = 22$ kΩ, $R_1 = R_a = 1.1$ kΩ, $R_L = 500$ Ω, and the Op Amp is a type 741 with dc supply voltages of ± 15 V, with what maximum voltage V_s can we drive this differential circuit and still not cause clipping of the output signal? What is the typical common-mode gain A_{cm}?

17. Over what range can the gain V_o/V_s in the circuit of Fig. 5-15 be adjusted?

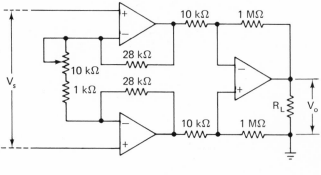

Fig. 5-15

18. If the 1-MΩ resistors and the 28-kΩ resistors are replaced with 10-kΩ values in the circuit of Fig. 5-15, over what range can its gain be adjusted?

19. What is the gain V_o/V_s of the circuit in Fig. 5-15 if its 1-kΩ resistor becomes open?

20. In the circuit of Fig. 5-15, if the 1-kΩ resistor is replaced with a short, to what theoretical maximum can this circuit's voltage gain be adjusted?

21. What is the gain V_o/V_s of the circuit in Fig. 5-16?

22. What is the gain V_o/V_s of the circuit in Fig. 5-16 if both 1-kΩ resistors are replaced with 2.2 kΩ values?

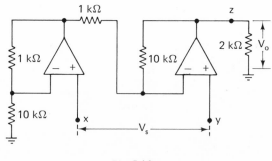

Fig. 5-16

23. In the circuit of Fig. 5-16, if the voltage at point x is $+5.5$ V to ground and the voltage at y is $+5.1$ V to ground, what is the voltage at point z with respect to ground, assuming that the output was initially nulled? What is the average dc common-mode input voltage?

24. In the circuit of Fig. 5-16, if the voltages at points x and y are -0.1 V and -0.5 V, respectively, what is the voltage at point z with respect to ground, and what is the dc common-mode input voltage? Assume that the output was initially nulled.

CHAPTER 6

THE OP AMP'S
BEHAVIOR AT
HIGHER FREQUENCIES

Some of the characteristics of the practical Op Amp are sensitive to changes in operating frequency. In this chapter, we will consider problems caused by the decrease of the open-loop gain (roll-off) at higher frequencies. We will also see how the gain vs frequency curve of an Op Amp can be altered, either by the circuit designer who uses the Op Amp or by the manufacturer. Other notable frequency-related problems such as reduced output-voltage swing capabilities, limited slew rates, and noise are discussed here too.

6-1 Gain and Phase Shift vs Frequency

Ideally, an Op Amp should have an infinite bandwidth. This means that, if its open-loop gain is 90 dB with dc signals, its gain should remain 90 dB through audio and on to high radio frequencies. The practical Op Amp's gain, however, decreases (rolls off) at higher frequencies as shown in Fig. 6-1. This gain roll-off is caused by capacitances within the Op Amp circuitry. The reactances of these capacitances decrease at higher frequencies, causing shunt signal current paths, and thus reducing the amount of signal available at the output terminal. Along with decreased gain at higher frequencies, there is an increased phase shift of the output signal with respect to the input. See Fig. 6-2. Normally at low frequencies, there is a 180° phase difference in the signals at the inverting input and output terminals. But at higher frequencies, the output signal lags by more than 180°, and this is called *phase shift*. Thus the Op Amp with the characteristics in Fig. 6-2 has practically no phase shift up to about 30 kHz. Beyond 30 kHz, the output signal starts

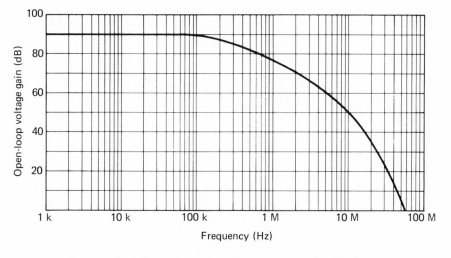

Fig. 6-1 Typical open-loop gain vs frequency curve of an Op Amp.

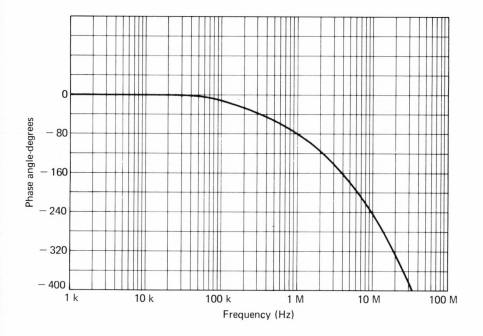

Fig. 6-2 Typical open-loop output-signal phase shift vs frequency.

to lag, and at 300 kHz it lags an additional 40°. This negative or lagging shift adds to the initially lagging 180°, causing an output signal lag of 220° compared with the signal applied to the inverting input. Similarly, at 1 MHz, the output lags the inverting input's signal by an additional 80°, or a 260° total. A serious problem caused by this kind of output-signal phase shift is discussed later in this chapter.

6-2   Bode Diagrams

The gain and phase shift curves are often approximated with a series of straight lines.* Such straight-line approximations of gain and phase shift vs frequency, where frequency is plotted on a logarithmic scale, are called *Bode diagrams* or *Bode plots*. A straight-line approximation of the curve in Fig. 6-1 is shown in Fig. 6-3. Note that, according to the approximated curve,

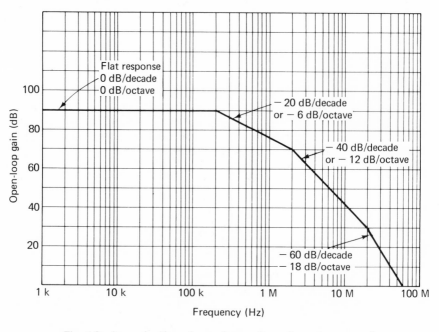

Fig. 6-3 Approximation of open-loop gain vs frequency curve of Fig. 6-1.

---

* The straight lines are linear sums of the asymptotes on the individual stage gain vs frequency curves.

this Op Amp's frequency response is flat from low frequencies (including dc) to 200 kHz. This means that its gain is constant from zero to 200 kHz, and therefore the bandwidth BW is about 200 kHz. Note that at higher frequencies, from 200 kHz to 2 MHz, the gain drops from 90 dB to 70 dB, which is at a -20 dB/decade* or -6 dB/octave rate. The negative sign refers to the negative (decreasing) slope of the curve. At frequencies from 2 MHz to 20 MHz, the roll-off is -40 dB/decade or -12 dB/octave. Above 20 MHz, the roll-off is -60 dB/decade or -18 dB/octave.

Because Op Amps are seldom used in open loop for amplification of signals, we must consider the effects of feedback on the Op Amp's frequency response. Note in Fig. 6-4 that if the Op Amp is wired to have a closed-loop gain $A_v = 10{,}000$ or 80 dB, the bandwidth is about 600 kHz, that is, by projecting to the right of 80 dB, we intersect the open-loop gain curve at a point above 600 kHz. This 80-dB projection and the open-loop curve intersect at a 20 dB/decade rate of closure. If the Op Amp is wired for a closed-loop gain $A_v = 1000$ or 60 dB, the bandwidth *appears* to be about 3.5 MHz. However, because the 60-dB projection and the open-loop curve

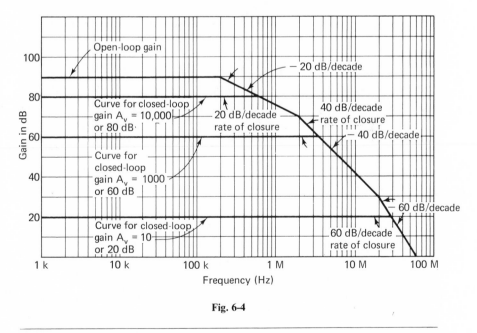

Fig. 6-4

---

* A -20 dB/decade gain roll-off means that the open-loop gain decreases 20 dB with a frequency increase by the factor 10, and is equivalent to a -6 dB/octave roll-off. Consequently, the open-loop gain decreases 6 dB with a frequency increase by the factor 2.

intersect at a 40 dB/decade rate of closure, the circuit is likely to be unstable and should not be used without modifications. The main reasons for circuit instability, which means that it might go into oscillation, are the following:

(1) In the −40 dB/decade, or more, roll-off region, the Op Amp's output signal phase shift is so large that it becomes, or approaches, an in-phase condition with the signal at the inverting input.

(2) With lower closed-loop gains, the feedback resistor R_F is relatively small compared to resistance R_1. See Figs. 3-5 and 3-10. This increases the amount of signal that is fed back to the inverting input.

The effect of the output signal becoming more in-phase with the signal normally at the inverting input, combined with the greater amount of this output being fed back, makes the Op Amp oscillate at higher frequencies when wired to have relatively low closed-loop gain. In other words, at higher frequencies and lower closed-loop gains, the feedback becomes significant and regenerative and thus meets the requirements of an oscillator.

Generally, the rate of closure between the closed-loop gain projection and the open-loop curve should not significantly exceed 20 dB/decade or 6 dB/octave for stable operation. Therefore the Op Amp with the characteristics in Fig. 6-4 is likely to be unstable if wired for gains below about 70 dB. Thus as mentioned before, if this Op Amp is wired for 60-dB gain, the rate of closure is 40 dB/decade, and the circuit will probably be unstable. Also as shown in Fig. 6-4, a gain of 20 dB causes a 60 dB/decade rate of closure which also means unstable operation. When unstable, the circuit can have unpredictable output signals even if no input signal is applied.

6-3   External Frequency Compensation

Some types of Op Amps are made to be used with externally connected compensating components—capacitors and sometimes resistors—if they are to be wired for relatively low closed-loop gains. These are called *uncompensated* or *tailored-response* Op Amps because the circuit designer must provide the compensation if it is required. The compensating components alter the open-loop gain characteristics so that the roll-off is about 20 dB/decade over a wide range of frequencies. For example, Fig. 6-5 shows some gain vs frequency curves that are obtained with various compensating components on a tailored-response Op Amp. In this case, if $C_1 = 500$ pF,

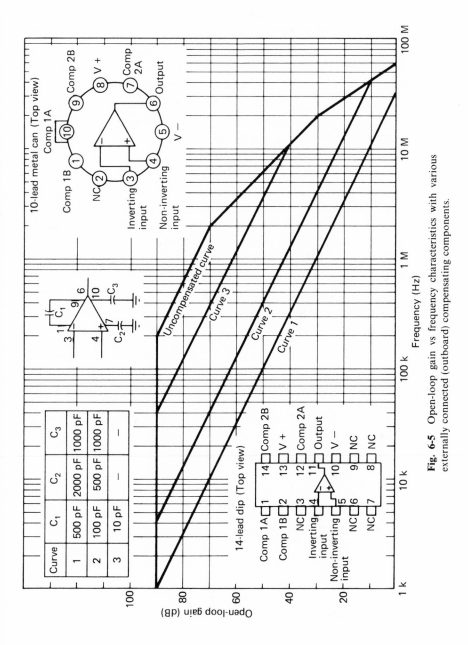

Fig. 6-5 Open-loop gain *vs* frequency characteristics with various externally connected (outboard) compensating components.

$C_2 = 2000 \, \text{pF}$, and $C_3 = 1000 \, \text{pF}$, then the gain vs frequency curve 1 applies according to the table, and the roll-off starts at about 1 kHz with a steady $-20 \, \text{dB/decade}$ rate. Thus if feedback components are now added to obtain an 80-dB closed-loop gain, the circuit's frequency response is flat from dc to about 3 kHz. This can be seen if we project to the right of 80 dB to curve 1. The intersection of this projection and curve 1 is at 3 kHz. Better bandwidth is obtained with curve 1 if a low closed-loop gain is used, that is, if $A_v = 100$ or 40 dB, the bandwidth increases to about 300 kHz. In either case, the rate of closure is 20 dB/decade, and the circuit is stable. Generally, if low gain is required, compensating components for curve 1 should be used. If high gain and relatively wide bandwidth are required, the compensating component for curve 3 should be used. Note in the table that if curve 3 is used, only one compensating component, $C_1 = 10 \, \text{pF}$, is required in this case. No compensating components need be used if this Op Amp is wired for more than 70 dB because the rate of closure never exceeds 20 dB/decade with such high gains. Note also on curve 3 that a closed-loop gain of less than 30 dB causes this gain vs frequency response curve (projection) to meet the open-loop gain curve at a point where the rate of closure is greater than 20 dB/decade; therefore the circuit is likely to be unstable.

EXAMPLE 6-1

If the data in Fig. 6-5 applies to the Op Amp in Fig. 6-6, what are this circuit's gain V_o/V_s and bandwidth with each of the following switch positions: I, II, and III? Is this circuit stable in each switch position?

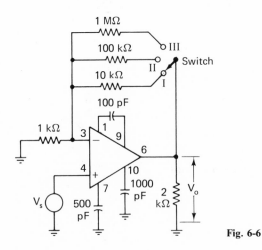

Fig. 6-6

Answer. This is a noninverting circuit and therefore its gain is

$$A_v = V_o/V_s \cong \frac{R_F}{R_1} + 1 \qquad (3\text{-}6)$$

The compensating capacitors give us the gain vs frequency curve 2. With the switch in position I,

$$A_v \cong \frac{10 \text{ k}\Omega}{1 \text{ k}\Omega} + 1 = 11$$

This is equivalent to about 20.8 dB or roughly 20 dB. Projecting to the right from 20 dB in Fig. 6-5, we intersect curve 2 at 13 MHz, which means that the bandwidth in this case is about 13 MHz.

With the switch in position II,

$$A_v \cong \frac{100 \text{ k}\Omega}{1 \text{ k}\Omega} + 1 = 101$$

which is equivalent to about 40 dB. Projecting to the right of 40 dB we intersect curve 2 at about 1.3 MHz, and therefore the bandwidth now is about 1.3 MHz.

With the switch in position III,

$$A_v \cong \frac{1 \text{ M}\Omega}{1 \text{ k}\Omega} + 1 = 1001$$

which is equivalent to about 60 dB. A projection to the right of 60 dB intersects the curve 2 above 130 kHz, which means that the bandwidth now is 130 kHz.

Since all three projections intersect curve 2 at a -20 dB/decade rate of closure, the circuit is stable regardless of the switch position.

EXAMPLE 6-2

If the 500-pF and 1000-pF capacitors are removed and the 100 pF is replaced with a 10-pF capacitor in the circuit of Fig. 6-6, what are the gain and bandwidth with each of the switch positions? Is this circuit stable in each switch position? Assume that the data in Fig. 6-5 applies to the Op Amp in this circuit.

Answers. According to the table in Fig. 6-5, with $C_1 = 10$ pF while C_2 and C_3 are removed, the gain vs frequency curve 3 applies.

With the switch in position I, $A_v \cong 11$, as before. A projection from 20 dB intersects curve 3 above 28 MHz approximately. However, since the rate of closure is -60 dB/decade at this intersection, the circuit is unstable and should not be used as an amplifier in this manner.

With the switch in position II, $A_v \cong 101$ or about 40 dB. A projection from 40 dB intersects curve 3 at 11 MHz approximately. This circuit is likely to be unstable too because the rate of closure is -40 dB/decade at the intersection.

In position III, $A_v = 1001$ or about 60 dB. Projecting to the right of 60 dB we intersect curve 3 at 1.1 MHz approximately. With a -20 dB/decade rate of closure at this intersection, we can expect a stable circuit with about a 1.1-MHz bandwidth.

6-4   Compensated Operational Amplifiers

Sometimes the relatively broad bandwidth of the uncompensated Op Amps is not needed. For example, in the instrumentation circuit shown in Fig. 5-8 of the previous chapter, the Op Amp is required to amplify relatively slow changing signals, and therefore it doesn't require good high-frequency response. In this and similar applications, internally compensated Op Amps can be used. They are sometimes simply called *compensated* Op Amps as shown in Appendix F1. Also, they are stable regardless of the closed-loop gain and without externally connected compensating components.

The type 741 Op Amp is compensated and has an open-loop gain vs frequency response as shown in Fig. 6-7. The IC of the 741 contains a 30-pF capacitance that internally shunts off signal current and thus reduces the available output signal at higher frequencies. This internal capacitance, which is an internal compensating component, causes the open-loop gain to roll off at a steady 20 dB/decade rate. Therefore, regardless of the closed-loop gain we use, the gain projection will intersect the open-loop gain curve at a -20 dB/decade rate of closure, and this assures us of a stable circuit.

The 741, along with some other types of compensated Op Amps, has a 1-MHz *gain-bandwidth product*. This means that the product of the coordinates, gain and frequency, of any point on the open-loop gain vs frequency curve is about 1 MHz.

Obviously, if the 741 Op Amp is wired for a closed-loop gain of 10^4 or 80 dB, its bandwidth is 100 Hertz as can be seen by projecting to the right from 10^4 to the curve in Fig. 6-7. If the closed-loop gain is lowered, say to

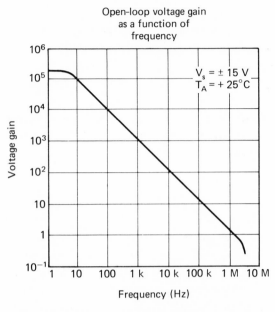

Open-loop voltage gain
as a function of
frequency

Fig. 6-7   Frequency response of the type 741 Op Amp.

10^2 or to 1, the bandwidth increases to 10 kHz or 1 MHz, respectively. The fact that the 741 has a 1-MHz bandwidth with unity gain explains why the 741 is listed with a *unity gain-bandwidth* of 1 MHz on typical specification sheets, as are many other compensated types.

EXAMPLE 6-3

If the Op Amp in the circuit of Fig. 6-8 is a type 741 and the circuit is required to amplify signals up to about 10 kHz, what maximum value of POT resistance can we use and still keep the circuit's response flat to 10 kHz?

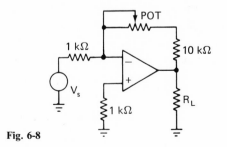

Fig. 6-8

Answer. The frequency response curve in Fig. 6-7 shows that the bandwidth is greater than 10 kHz if the closed-loop gain is kept under 100. Therefore, we want to keep the ratio R_F/R_1 under 100. Thus since

$$\frac{R_F}{R_1} < 100 \quad \text{then} \quad R_F < 100R_1 = 100(1 \text{ k}\Omega) = 100 \text{ k}\Omega$$

If we must keep the total feedback resistance R_F under 100 kΩ, then the POT's maximum resistance is 90 kΩ. Note the 10-kΩ fixed resistor in series with it.

6-5   Slew Rate

An Op Amp's slew rate is related to its frequency response. Generally, we can expect Op Amps with wider bandwidths to have higher (better) slew rates. The slew rate, as mentioned in Chapter 2, is the rate of output voltage change caused by a step input voltage and is usually measured in volts per microsecond. An ideal slew rate is infinite, which means that the Op Amp's output voltage should change instantly in response to an input step voltage. Practical IC Op Amps have specified slew rates from 0.1 V/μs to about 100 V/μs which are measured in special circumstances. Some hybrid* Op Amps have slew rates on the order of 1000 V/μs. Unless otherwise specified, the slew rate listed in a data sheet was probably measured with unity gain and open load. Often the slew rate improves with higher gains and dc supply voltages. See Fig. 6-9.

A less than ideal slew rate causes distortion especially noticeable with nonsinusoidal waveforms at higher frequencies. For example, ignoring overshoot, Fig. 6-10 shows typical output waveforms of a voltage follower with square waves of various frequencies applied. In this case, the slew rate is about 1 V/μs. When the input frequency is 100 Hz with the waveform in Fig. 6-10b, the output has the waveform shown in part c of the figure. Similarly, when the input is 10 kHz or 1 MHz with the waveform shown in b, the output has the waveform in Fig. 6-10d or e, respectively. Obviously, due to the limited slew rate, the square-wave input is distorted into a sawtooth at higher frequencies. Where high rates of output voltage change are required (large dV_o/dt), *high-speed* Op Amps are used such as the 715 and

---

* Hybrids, as opposed to monolithics (made on one chip), might contain two or more ICs or ICs and discrete components.

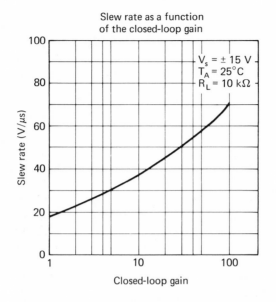

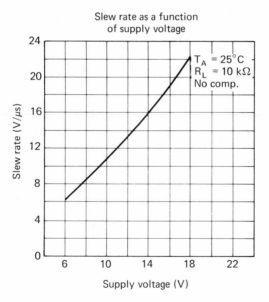

Fig. 6-9 Slew rate of the type 715 Op Amp increases with gain and supply voltage.

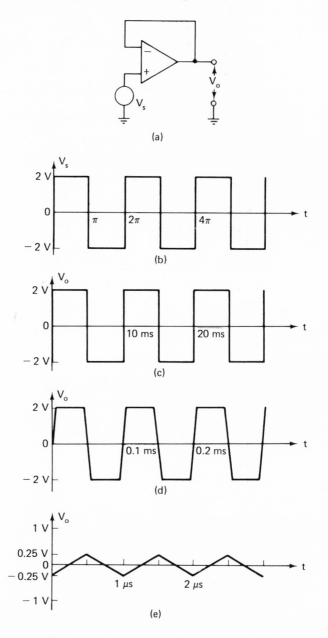

Fig. 6-10 Voltage follower with an Op Amp having a slew rate of 1 V/μs and square-wave voltage applied. (c) Output when the frequency of V_s is 100 Hz. (d) Output when the frequency of V_s is 10 kHz. (e) output when the frequency of V_s is 1 MHz.

the 776 types. See Appendix F2. The 101A's slew rate is improved when wired with *feedforward compensation* as shown in Fig. 6-11. The type 108

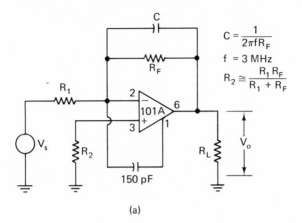

$$C = \frac{1}{2\pi f R_F}$$

$$f = 3 \text{ MHz}$$

$$R_2 \cong \frac{R_1 R_F}{R_1 + R_F}$$

(a)

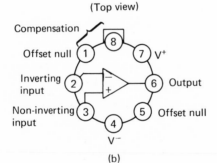

(b)

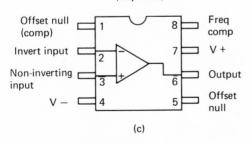

(c)

Fig. 6-11 (a) A type 101A Op Amp connected as feedforward compensated amplifier for higher slew rate. (b) An 8-pin metal can. (c) An 8-pin minidip (dual in-line package).

Op Amp in Fig. 6-12 is similarly connected with feedforward compensation to improve its slew rate and bandwidth. In this case the compensating components not only broaden the bandwidth but also cause a 20 dB/decade open-loop gain vs frequency roll-off and thus provide stable operation.

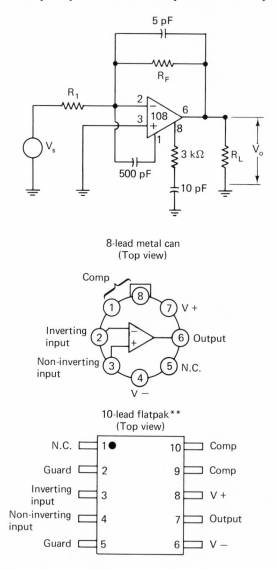

Fig. 6-12 Feedforward compensation. The compensating components are the capacitors and the 3-kΩ resistor.

6-6   Output Swing vs Frequency

An important consideration is an Op Amp's peak-to-peak output signal capability at higher frequencies. Generally, this capability decreases with higher frequencies, as shown in Fig. 6-13. In Fig. 6-13a, the type 777 Op Amp

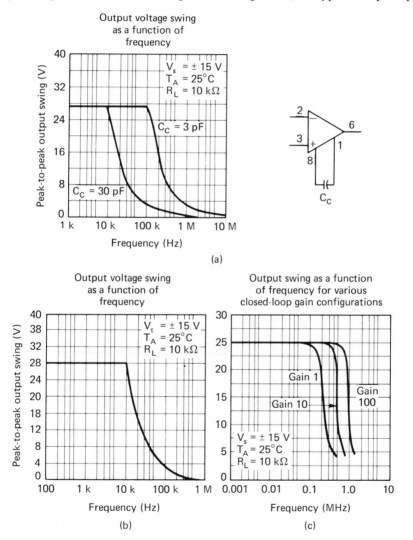

Fig. 6-13 Peak-to-peak output voltage capability vs frequency of the (a) uncompensated type 777 Op Amp; (b) compensated type 741 Op Amp; and (c) uncompensated high-speed type 715 Op Amp.

characteristics are shown for two different compensating capacitors C_c. Note that if $C_c = 30$ pF is used, the peak-to-peak output capability decreases sharply at operating frequencies above 10 kHz. On the other hand, when $C_c = 3$ pF is wired externally (outboard), the bandwidth increases, as it did with smaller compensating capacitors shown in Fig. 6-5. Also, the peak-to-peak output capability does not drop until frequencies on the order of 100 kHz or more are applied.

The output vs frequency curve in Fig. 6-13b is for the type 741 Op Amp, which is internally compensated. Apparently then, internal compensation not only limits bandwidth, but it also limits the peak-to-peak output capability. We must keep in mind, though, that internal compensation has outstanding advantages for instrumentation and low-frequency work. For example, it simplifies our work by not forcing our attention to externally wired compensation and by assuring us of stable operation. As would be expected, faster Op Amps, intended for broadband applications, have good peak-to-peak output capability into higher frequencies. See Fig. 6-13c.

EXAMPLE 6-4

If the Op Amp in the circuit of Fig. 6-8 is a type 741 with dc source voltages of ± 15 V, $R_L = 10$ kΩ, and the POT adjusted to zero, with which of the following audio-frequency input signals will this circuit clip the output signals? (Assume that the Op Amp was initially nulled and that the input signals have no dc component.)

(a) 1 Hz to 10 kHz with peaks up to ± 1 V;
(b) 1 Hz to 100 kHz with peaks up to ± 1 V;
(c) 1 Hz to 100 kHz with peaks up to ± 10 mV.

Answer. With the POT resistance 0 Ω, the total feedback resistance is 10 kΩ. Since this is an inverting amplifier, its gain $A_v = -R_F/R_1 = -10$ k$\Omega/1$ k$\Omega = -10$. With this gain, the 741's bandwidth is about 100 kHz. See Fig. 6-7. All three signals are within the bandwidth capability of this circuit. Therefore, we can focus our attention to the peak-to-peak output capability vs frequency.

(a) With the 1-Hz to 10-kHz input signal peaking to ± 1 V, the output will peak to ± 10 V. Since the 741 can handle from 28 V peak-to-peak (14 V peak) up to 10-kHz, no clipping occurs.

(b) With the 1-Hz to 100-kHz input signal peaking up to ± 1 V, the output *attempts* to peak at ± 10 V over this frequency range. However, according to the curve in Fig. 6-13b, this output is clipped with signal frequencies above 15 kHz approximately. Projecting to the right of 20 V peak-to-peak (10 V peak), we intersect the curve at roughly 15 kHz. Thus input signals with peaks of 1 V are clipped more or less, depending on how far above 15 kHz we are working.

(c) With the 1-Hz to 100-kHz input signal peaking up to ± 10 mV, the output peaks at ± 100 mV. No clipping occurs because the 741's peak-to-peak output capability is about 2 V or 1 V peak up to 100 kHz. This capability is well over the ± 100-mV peaks we intend to get.

6-7   Noise

In Chapter 5 we referred to induced 60-Hz hum as noise. Indeed, any unwanted signal mixing with desired signals is noise. Induced noise is not limited to the 60 Hz from nearby electrical power equipment. It often originates in other man-made systems such as switch arcing, motor brush sparking, and ignition systems. Natural phenomena such as lightning also cause induced noise. Induced noise voltages, as shown in Chapter 5, can be made common mode and thus can be significantly reduced relative to the desired signals at the load of an Op Amp.

The term *noise* is also commonly used to describe ac random voltages and currents generated within conductors and semiconductors. Such noise, associated with Op Amps and with amplifiers in general, limit their signal sensitivity. If very weak signals are to be amplified, very high closed-loop gain must be used to bring the signals up to useful levels. With a very high gain, however, the noise is amplified along with the signals to the point where nearly as much noise as signal appears at the output. If fed to a speaker, random noise causes a hissing, frying sound.

There are three main types of noise phenomena associated with Op Amps and with solid-state amplifiers in general. These are: *shot* or *Schottky* noise, *thermal* or *Johnson* noise, and *flicker* or $1/f$ noise.

Thermal noise is caused by the random motion of charge carriers within a conductor which generate noise voltages V_n within it. The rms value of this thermally generated noise voltage V_n can be predicted with the equation

$$V_n = \sqrt{4KTR(BW)} \qquad (6\text{-}1)$$

where: K is Boltzmann's constant; 1.38×10^{-23} joules/°K,

T is the temperature in degrees Kelvin—the Celsius temperature plus 273°,

R is the resistance of the conductor in question, and

BW is the bandwidth in Hertz.

Examining this equation, we see that thermal noise increases with higher temperatures, larger resistances, and wider bandwidths.

Although a constant average dc current may be maintained in a semiconductor, it will have random variations. These variations have an rms value referred to as noise current I_n. Noise generated in this manner is called *shot noise*. Its rms value can be predicted with the equation

$$I_n = \sqrt{2qI_{dc}(BW)} \qquad (6\text{-}2)$$

where: q is the charge of an electron: 1.6×10^{-19} coulombs,

I_{dc} is the average dc current in the semiconductor, and

BW is the bandwidth.

Here again, wider bandwidths generate more noise, which is to say, we can expect less noise in narrow-bandwidth amplifiers. Of course, the noise current I_n, flowing through a resistance R, will generate noise voltage RI_n.

In addition to shot noise, semiconductors have low-frequency noise called *flicker* or $1/f$ noise. The term $1/f$ describes the inverse nature of this noise with respect to frequency, that is, the amount of flicker noise is greater with lower values of frequency f.

6-8   Equivalent Input Noise

An Op Amp contains many active and passive components that generate and add noise to its output. These noise sources can be represented by voltage and current noise generators at the input of the amplifier's equivalent circuit, as shown in Fig. 6-14. This equivalent is sometimes called an *amplifier noise model*. The net effect of these input noise generators is an *equivalent input-noise voltage* V_{ni}. It is this equivalent input-noise V_{ni} that is amplified by the stage gain, and with it we can predict the output noise V_{no} that a given Op Amp stage might have.

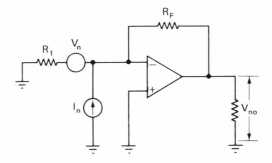

Fig. 6-14 An amplifier's inherent noise voltage and current can be shown as voltage and current generators at its inputs.

If we neglect the thermal noise of the amplifier's signal source, the equivalent input noise V_{ni} is related to the amplifier's specified input noise voltage V_n and noise current I_n by the following equation:

$$V_{ni}^2 \cong V_n^2 + (R_E I_n)^2 \qquad (6\text{-}3)$$

where: V_{ni}^2 is the mean-square equivalent input noise voltage,
$\quad\;\;\; V_n^2$ is the specified mean-square input noise voltage,
$\quad\;\;\; I_n^2$ is the specified mean-square input noise current, and
$\quad\;\;\; R_E$ is the parallel equivalent of R_1 and R_F.

It can be shown that equivalent input noise voltage V_{ni} sees the Op Amp circuit as a noninverting type as shown in Fig. 6-15. Broadband input noise vs source resistance curves are sometimes shown as in Fig. 6-15b, and the general-purpose Op Amp's specified input noise voltage V_n and input noise current I_n vs frequency characteristics are shown in Fig. 6-15c and d. Larger source resistances appreciably add to thermally generated noise as shown in part b of the figure. The $1/f$ phenomenon adds low-frequency noise as shown in Fig. 6-15c and d.

The output noise voltage V_{no} is a function of the effective input noise voltage V_{ni} and is approximated with the equation

$$V_{no}^2 \cong (A_v V_{ni})^2 \qquad (6\text{-}4a)$$

or

$$V_{no} \cong A_v \sqrt{V_n^2 + (R_F I_n)^2} \qquad (6\text{-}4b)$$

where A_v is the closed-loop gain.

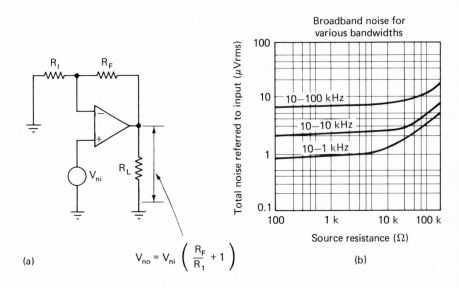

(a)

$$V_{no} = V_{ni} \left(\frac{R_F}{R_1} + 1 \right)$$

(b)

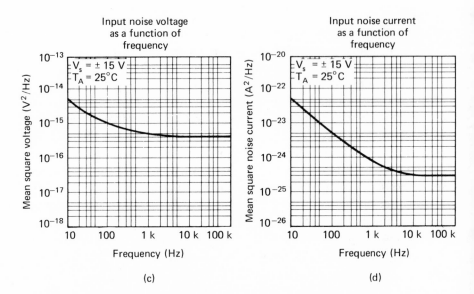

(c)

(d)

Fig. 6-15 (a) The total effective input noise is amplified by the closed-loop gain of the amplifier seeing it as a noninverting type. (b) Typical noise levels higher with wider bandwidths and with larger signal source resistances. (c) The rms noise voltages squared typically developed at various frequencies. (d) The rms noise currents squared typically developed at various frequencies.

REVIEW QUESTIONS

1. Referring to the open-loop gain of an Op Amp, what does the term *roll-off* mean?

2. What is a Bode diagram?

3. What problem can occur with an amplifier having a steep gain roll-off?

4. How can the steepness of the gain roll-off deliberately be limited?

5. Why are low closed-loop gains avoided with uncompensated Op Amps?

6. What advantage does the $-20 \, \text{dB/decade}$ roll-off of the internally compensated Op Amp have?

7. What disadvantage does the internally compensated Op Amp have compared to the externally compensated types?

8. If an Op Amp is specified as having a *tailored response*, what does this mean?

9. If an Op Amp is specified as being *compensated*, what does this mean?

10. If the closed-loop gain of an Op Amp is increased by increasing its feedback resistance R_F, what effect will this have on its bandwidth?

11. If an Op Amp's gain-bandwidth product is 2 MHz, what is its bandwidth when connected to work as a voltage follower?

12. What is an Op Amp's slew rate?

13. What effect does the operating frequency have on the maximum unclipped output signal capability of an Op Amp?

14. When compensating components are added externally on an Op Amp, what effect do they have on the bandwidth?

15. Name the three types of noise associated with conductors and semiconductors.

16. Generally, at what frequencies is flicker noise a greater problem?

17. What effect do large externally wired resistances have on the noise level at the output of an Op Amp?

PROBLEMS

Refer to Fig. 6-16a. The Op Amp will work on one of the various open-loop vs frequency curves shown in Fig. 6-16c, depending on the values of the externally wired compensating components actually used.

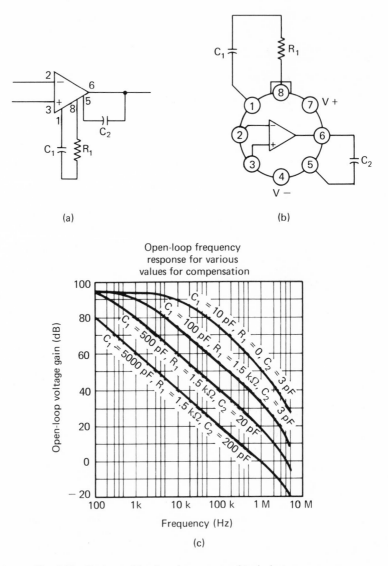

(a)   (b)

(c)

Fig. 6-16   Op Amp with tailored response, and typical response curves.

1. For the Op Amp and characteristics of Fig. 6-16, what is the gain-band-width product if $C_1 = 5000\,\text{pF}$, $R_1 = 1.5\,\text{k}\Omega$, and $C_2 = 200\,\text{pF}$?

2. For the Op Amp and characteristics of Fig. 6-16, what is the gain-bandwidth product in the frequency range from 1 kHz to 1 MHz if $C_1 = 500\,\text{pF}$, $R_1 = 1.5\,\text{k}\Omega$, and $C_2 = 20\,\text{pF}$?

3. If we are after as wide a bandwidth as possible, with a gain on the order of a few thousands, what compensating components would you recommend for the Op Amp and characteristics of Fig. 6-16?

4. In Fig. 6-16, what compensating component values are likely to make the circuit unstable if the circuit is wired as a voltage follower?

5. If the Op Amp in Fig. 6-17 has the characteristics given in Fig. 6-16, what is the circuit's bandwidth when the 100-kΩ POT is adjusted to 0 Ω?

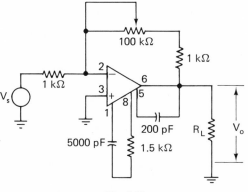

Fig. 6-17

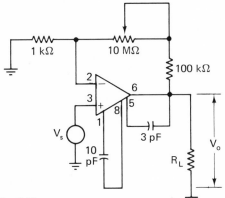

Fig. 6-18

6. If the Op Amp in Fig. 6-17 has the characteristics given in Fig. 6-16, what is the circuit's bandwidth when the 100-kΩ POT is adjusted to maximum?

7. What is the bandwidth of the circuit in Fig. 6-18 if its 10-MΩ POT is adjusted for maximum resistance and if its Op Amp has the characteristics given in Fig. 6-16?

8. What is the bandwidth of the circuit in Fig. 6-18 if its 10-MΩ POT is adjusted to 1 MΩ and if its Op Amp has the characteristics given in Fig. 6-16?

9. In the circuit of Fig. 6-19a, what are its approximate gain and bandwidth if the 1-kΩ POT is adjusted to 0 Ω and if the Op Amp has the gain vs frequency curve in b of the figure?

10. If the 1-kΩ POT in the circuit of Fig. 6-19a is adjusted for maximum resistance and if the Op Amp has a gain-bandwidth product of 1 MHz, what are this circuit's gain and bandwidth?

11. If the Op Amp in the circuit of Fig. 6-17 has the characteristics in Fig. 6-16c, what is the rate of closure between the closed-loop and the open-loop gains vs frequency curves regardless of the adjustment of the 100-kΩ POT?

12. In Fig. 6-19a, if the Op Amp has the gain vs frequency curve in b of the figure, what is the rate of closure between the closed-loop and the open-loop gains vs frequency curves with most adjustments of the 1-kΩ POT?

13. The Op Amp circuit in Fig. 6-20b has the characteristics shown in Fig. 6-20a. Its input signal V_s has variable frequency and often peaks up to 80 mV. If $R_F = 100$ kΩ, beyond what applied frequency is the output signal V_o likely to be clipped? Assume that the circuit was initially nulled.

14. Refer to the Op Amp circuit and characteristics in Fig. 6-20. What maximum peak-to-peak voltage value of V_s can we apply and not clip the output signal V_0 if $R_F = 200$ kΩ and the amplifier must have a flat response up to 100 kHz?

15. Refer to the Op Amp and its characteristics in Fig. 6-21. How much common-mode voltage can we expect across the load R_L if the input common-mode voltage $V_{cm} = 2$ mV rms at 60 Hz?

16. Refer to the Op Amp and its characteristics in Fig. 6-21. How much common-mode voltage can we expect across the load R_L if the input common-mode voltage $V_{cm} = 2$ mV at 100 kHz?

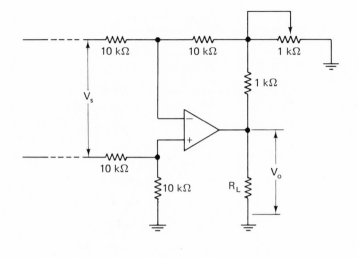

(a)

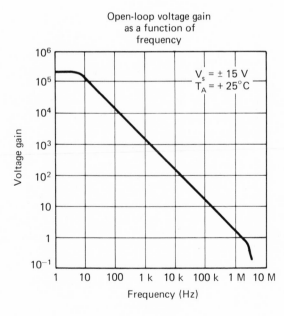

(b)

Fig. 6-19

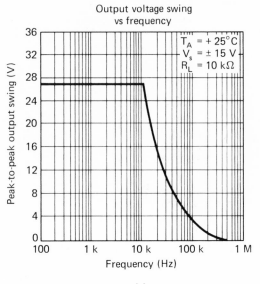

(a)

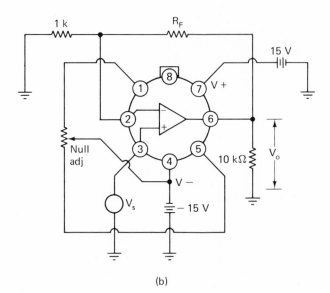

(b)

Fig. 6-20

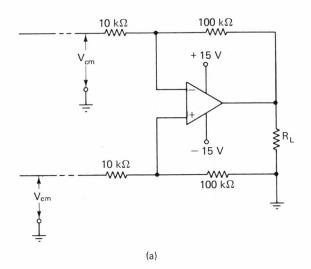

(a)

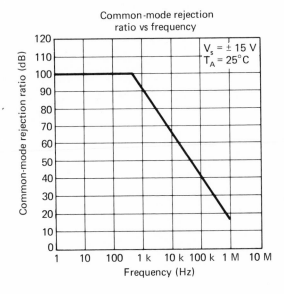

(b)

Fig. 6-21

17. Refer to Fig. 6-10. If the Op Amp circuit in part a has a slew rate of 0.5 V/μs and has the input signal V_s waveform in b of the figure applied, sketch the output waveform and indicate its peak-to-peak value if the frequency of V_s is 62.5 kHz. Assume that the frequency response is flat to 1 MHz.

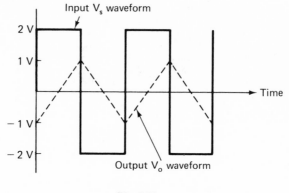

Fig. 6-22

18. If a voltage-follower circuit has the square-wave input and sawtooth output waveforms of Fig. 6-22, what is the Op Amp's slew rate? The square-wave frequency $f = 1$ MHz.

19. If the total effective input noise voltage is 5 μV rms in the circuit of Fig. 6-21, what is this circuit's output noise voltage?

CHAPTER 7

PRACTICAL
CONSIDERATIONS

In the preceding chapters we learned what an Op Amp's characteristics should be ideally and what they are in practice. In this chapter, we will see that many of the practical Op Amp characteristics, as listed on its specification sheets, are not constant and tend to drift with drifting ambient conditions. The effect of such drifting characteristics is usually most noticeable at the output of the Op Amp, and we will see methods of predicting output voltage changes vs temperature and dc voltage source changes. In addition, we will see the characteristics of some special-purpose Op Amps and protecting techniques that are sometimes necessary in Op Amp circuits.

7-1   Offset Voltage vs Power Supply
Voltage

Because the Op Amp is capable of amplifying dc voltages, it is inherently sensitive to changes in its own dc supply voltages: the $+V$ and $-V$ sources. With practical Op Amps, if the dc supply voltages change due to poor regulation, the dc offset voltages will change too. Similarly, if the supply voltages are poorly filtered and vary at some ripple frequency, the offset voltages in the Op Amp will vary at the same frequency.

The sensitivity of an Op Amp to variations in the supply voltages is usually specified in a variety of somewhat equivalent terms such as the *power supply rejection ratio*, the *supply voltage rejection ratio*, the *power supply sensitivity*, and the *supply voltage sensitivity* to name a few. These are specified in decibels or microvolts per volt. For example,

$$\text{Power Supply Rejection Ratio, } PSRR, \text{ in decibels} = 20 \log \frac{\Delta V}{\Delta V_{io}} \qquad (7\text{-}1)$$

129

or

$$\text{Power Supply Sensitivity, } S, \text{ in } \mu V/V = \frac{\Delta V_{io}}{\Delta V} \qquad (7\text{-}2)$$

where: ΔV is the change in the power supply voltage, and

ΔV_{io} is the resulting change in input offset voltage.

The significance of these parameters is more apparent if we recall that the input offset voltage V_{io} is the voltage required across the differential inputs to null the output of the Op Amp with no feedback (open loop). We also learned in Chapter 4 that, with feedback, the input offset voltage sees the circuit as a noninverting amplifier (see Fig. 4-5) and will cause an output offset voltage V_{oo} if the circuit has not been nulled. This output offset is therefore the stage gain times the input offset, that is,

$$V_{oo} = A_v V_{io} \qquad (4\text{-}1)$$

or

$$V_{oo} \cong \left(\frac{R_F}{R_1} + 1\right) V_{io}$$

Apparently then, any input offset voltage *change* ΔV_{io}, caused by a power supply voltage *change* ΔV, will be amplified and cause an output offset voltage *change* ΔV_{oo} that is larger by the stage gain. Equation (4-1) can therefore be modified to

$$\Delta V_{oo} \cong \left(\frac{R_F}{R_1} + 1\right) \Delta V_{io} \qquad (7\text{-}3)$$

Changes in the input offset voltage ΔV_{io} can be determined if we rearrange Eq. (7-1) or (7-2). If the power supply rejection ratio in dB is given, then

$$\Delta V_{io} = \frac{\Delta V}{\text{antilog}\,(PSRR(\text{dB})/20)} \qquad (7\text{-}1a)$$

where the antilog of $(PSRR(\text{dB})/20)$ can quickly be estimated with the table in Fig. 5-4. If a sensitivity factor S is given, then

$$\Delta V_{io} = S(\Delta V) \qquad (7\text{-}2a)$$

Since the change in supply voltage ΔV can be due to poor regulation *or* poor filtering, then ΔV can be the ripple voltage V_r riding on the total supply voltage. This means that, with poor filtering, $\Delta V = V_r$ and the input offset voltage varies (changes) at the ripple frequency.

Supply voltage changes ΔV are easily measured across the $+V$ and $-V$ terminals as shown in Fig. 7-1. The meter M can be a voltmeter or an oscilloscope that is capable of measuring dc voltages and will measure ΔV caused by poor regulation or ΔV caused by poor filtering.

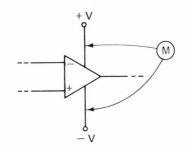

Fig. 7-1 Meter M reads the total dc supply voltage.

EXAMPLE 7-1

Referring to Fig. 7-2, suppose the circuit is nulled when the voltage across terminals $+V$ and $-V$ measures 30 V dc and that, due to poor regulation, this dc voltage drifts with time from 28 V to 32 V. Also suppose that, due to poor filtering, 2.5 mV rms ac ripple is also measured across the terminals $+V$ and $-V$. While the signal source $V_s = 0$ V, what drift occurs in the output offset voltage and how much ripple voltage can we expect across the load R_L if:
(a) the Op Amp's $PSRR = 80$ dB, or
(b) the Op Amp's power supply sensitivity $S = 150 \, \mu V/V$?

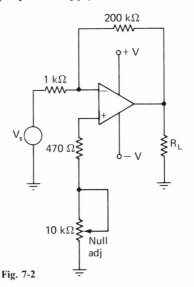

Fig. 7-2

Answer. Due to poor regulation, the dc supply can drift 2 V in either direction from 30 V; therefore, let $\Delta V = 2$ V. Due to poor filtering, $\Delta V = 2.5$ V rms.

(a) With the *PSRR* = 80 dB, which is equivalent to a voltage ratio of 10,000 (see Fig. 5-4), then due to the drift $\Delta V = 2$ V

$$\Delta V_{io} = \frac{2 \text{ V}}{\text{antilog } (80 \text{ dB}/20)} = \frac{2 \text{ V}}{10,000} = 0.2 \text{ mV} \qquad (7\text{-}1a)$$

and the resulting output offset drift is

$$\Delta V_{oo} \cong \left(\frac{R_F}{R_1} + 1\right)\Delta V_{io} = 201(0.2 \text{ mV}) = 40.2 \text{ mV} \qquad (7\text{-}3)$$

Due to the ripple $\Delta V = 2.5$ mV rms,

$$\Delta V_{io} = \frac{2.5 \text{ mV}}{10,000} = 0.25 \text{ } \mu\text{V rms} \qquad (7\text{-}1a)$$

and the output ripple voltage is

$$V_{o(\text{ripple})} \cong \left(\frac{R_F}{R_1} + 1\right)\Delta V_{io} = 201(0.25 \text{ } \mu\text{V}) = 50.25 \text{ } \mu\text{V rms} \quad (7\text{-}3)$$

(b) With the sensitivity factor $S = 150 \text{ } \mu\text{V/V}$, which is equivalent to 150×10^{-6}, and a dc supply drift $\Delta V = 2$ V,

$$\Delta V_{io} = S(\Delta V) = 150 \times 10^{-6}(2 \text{ V}) = 300 \text{ } \mu\text{V} = 0.3 \text{ mV} \quad (7\text{-}1b)$$

and the resulting output offset drift

$$V_{oo} \cong 201(0.3 \text{ mV}) = 60.3 \text{ mV} \qquad (7\text{-}3)$$

Due to the ripple $\Delta V = 2.5$ mV rms,

$$\Delta V_{io} = 150 \times 10^{-6}(2.5 \text{ mV}) = 0.375 \text{ } \mu\text{V rms} \qquad (7\text{-}1b)$$

and this causes an output ripple voltage of

$$V_{o(\text{ripple})} \cong 201(0.375 \text{ } \mu\text{V}) \cong 75.4 \text{ } \mu\text{V rms} \qquad (7\text{-}3)$$

Sometimes a drift in input bias current vs dc supply voltage is specified on manufacturers' data sheets in terms of picoamperes per volt. For example, an input bias current vs dc supply characteristic might be specified as ± 10 pA/V, which means that the input bias current might either increase or decrease by as much as 10 pA for every one-volt change in the dc supply as measured across the $+V$ and $-V$ terminals. Any drift in the input bias current will tend to cause a drift in the output offset, which is discussed in more detail in the next section.

7-2   Offset Voltage vs Temperature

The typical and maximum values of input bias current I_B and input offset current I_{io}, as specified in manufacturers' data sheets, are usually values measured at 25°C, which is about room temperature. We usually cannot depend on these specified currents holding to steady values if the temperature of the Op Amp drifts. Typical I_B vs temperature and I_{io} vs temperature curves for a general-purpose IC Op Amp are shown in Fig. 7-3. It is interesting to note that these bias and offset currents increase with decreased temperatures, which is generally the case with transistor-input IC Op Amps. The input bias current rises with higher temperatures in FET-input types of Op Amps. In either case, temperature changes cause bias and offset current changes, resulting in a drift in the output offset voltage. High-performance Op Amps are available that feature extremely low drift with temperature changes. As would be expected, their costs are higher than for the general-purpose types. The point is, we should be aware of the economical IC Op Amp's tendency to drift. If the drift is excessive for the application we have in mind, we can then consider the higher-performance types.

We concern ourselves with either the drift in input bias current ΔI_B or with the drift in the input offset current ΔI_{io}, depending on how the circuit is wired. Generally, if an Op Amp's dc resistance to ground, as seen from its noninverting input 2, is negligible compared to the dc resistance to ground, seen from the inverting input 1, we can closely estimate the circuit's change or drift in output offset voltage with the equation

$$\Delta V_{oo} \cong R_F \Delta I_B \qquad (7\text{-}4)$$

where: R_F is the feedback resistance across the inverting input 1 and the output, and

ΔI_B is the change or drift in the input bias current.

If the dc resistances to ground as seen from both the inverting and non-inverting inputs are equal, the change or drift in the output offset voltage can be closely estimated with the equation

$$\Delta V_{oo} \cong R_F \Delta I_{io} \qquad (7\text{-}5)$$

where ΔI_{io} is the change or drift in the input offset current.

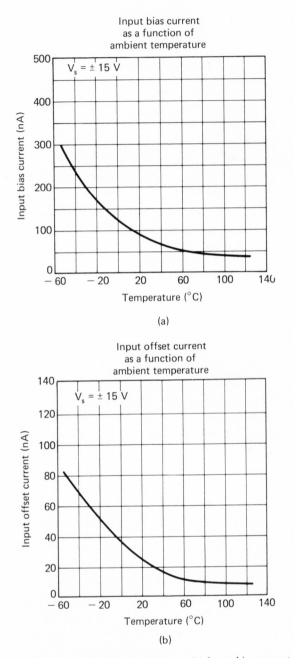

Fig. 7-3   General-purpose Op Amp curves: (a) input bias current I_B vs temperature; (b) input offset current I_{io} vs temperature.

For example, in Fig. 7-4a, the noninverting input 2 of the circuit is grounded; therefore, Eq. (7-4) is used to predict the drift in output offset ΔV_{oo}. Equation (7-4) also applies if R_2 is much smaller than $R_1 + R_s$ in the circuit of Fig. 7-4b and if R_s is much smaller than R_1 in the circuit in c of the figure.

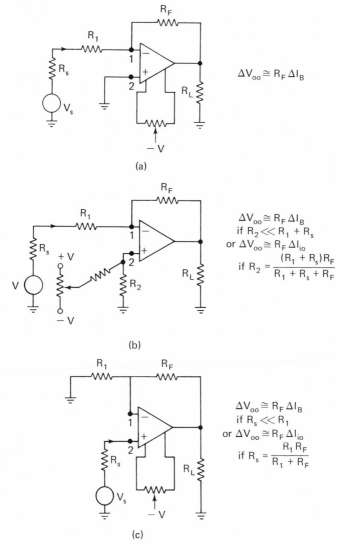

$$\Delta V_{oo} \cong R_F \Delta I_B$$

(a)

$$\Delta V_{oo} \cong R_F \Delta I_B$$
$$\text{if } R_2 \ll R_1 + R_s$$
$$\text{or } \Delta V_{oo} \cong R_F \Delta I_{io}$$
$$\text{if } R_2 = \frac{(R_1 + R_s)R_F}{R_1 + R_s + R_F}$$

(b)

$$\Delta V_{oo} \cong R_F \Delta I_B$$
$$\text{if } R_s \ll R_1$$
$$\text{or } \Delta V_{oo} \cong R_F \Delta I_{io}$$
$$\text{if } R_s = \frac{R_1 R_F}{R_1 + R_F}$$

(c)

Fig. 7-4   There is a change in output offset voltage ΔV_{oo} if a change in the input bias current ΔI_B occurs.

On the other hand, if

$$R_2 = \frac{(R_1 + R_s)R_F}{R_1 + R_s + R_F} \qquad (7\text{-}6)$$

in the circuit of Fig. 7-4b, and if

$$R_s = \frac{R_1 R_F}{R_1 + R_F} \qquad (7\text{-}7)$$

in the circuit of Fig. 7-4c, the dc resistances to ground seen looking from both inputs of each circuit are equal, and Eq. (7-5) applies.

Since ΔI_{io} is much smaller than the ΔI_B of a given Op Amp, designing for equal dc resistances to ground reduces an Op Amp circuit's drift. All the circuits of Fig. 7-5 are low-drift (stable) designs, provided their component values are properly selected. If R_2 is selected with Eq. (7-6) in the circuit of Fig. 7-5a, and if $R_a = R_1$ and $R_b = R_F$ in the circuit b of the figure, and if

$$R_2 = R_s - \frac{R_1 R_F}{R_1 + R_F} \qquad (7\text{-}8)$$

in the circuit of Fig. 7-5c, the output offset drift ΔV_{oo} is closely related to the offset current drift ΔI_{io} rather than bias current drift ΔI_B which make these circuits less prone to drift with temperature changes.

EXAMPLE 7-2

(a) The Op Amp in the circuit of Fig. 7-4b has the characteristics of Fig. 7-3 and is nulled at 25°C. What output offset is across the load R_L at 60°C if $R_s = 0\,\Omega$, $R_1 = 10\,\text{k}\Omega$, $R_F = 1\,\text{M}\Omega$, and $R_2 = 100\,\Omega$?

(b) The Op Amp in the circuit of Fig. 7-5b has the characteristics shown in Fig. 7-3 and is nulled at 25°C. What output offset is across the load R_L at 60°C if $R_a = R_1 = 10\,k\Omega$ and $R_b = R_F = 1\,\text{M}\Omega$?

Answer. (a) Since $R_2 \ll R_1 + R_s$, Eq. (7-4) and the curve of Fig. 7-3a apply. On this curve, $I_B \cong 80\,\text{nA}$ at 25°C, but it drops to 50 nA at 60°C. Thus, $\Delta I_B \cong 80 - 50 = 30\,\text{nA}$. This causes a

$$\Delta V_{oo} \cong R_F \Delta I_B \cong 1\,\text{M}\Omega(30\,\text{nA}) = 30\,\text{mV}$$

which means that the output voltage V_o drifts from its initial 0 V at 25°C to 30 mV as the temperature rises to 60°C.

(b) In this case, since $R_a = R_1$ and $R_b = R_F$, Eq. (7-5) and the curve in Fig. 7-3b apply. On this curve, $I_{io} \cong 25$ nA at 25°C, and it decreases to about 10 nA at 60°C. Therefore, with the $\Delta T = 60 - 25 = 35$°C,

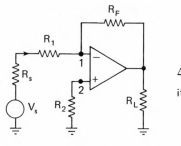

$$\Delta V_{oo} \cong R_F \Delta I_{io}$$
$$\text{if } R_2 = \frac{(R_1 + R_s)R_F}{R_1 + R_s + R_F}$$

(a)

$$\Delta V_{oo} \cong R_F \Delta I_{io}$$
$$\text{if } R_a = R_1$$
$$\text{and } R_b = R_F$$

(b)

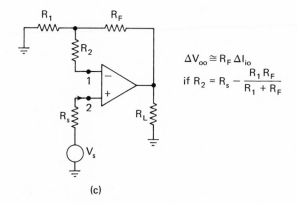

$$\Delta V_{oo} \cong R_F \Delta I_{io}$$
$$\text{if } R_2 = R_s - \frac{R_1 R_F}{R_1 + R_F}$$

(c)

Fig. 7-5 In stabilized circuits (R_2 properly selected), there is a change in the output offset voltage ΔV_{oo} if a change in the input offset current ΔI_{io} occurs.

the $\Delta I_{io} \cong 25 - 10 = 15\,\text{nA}$. The resulting change in output offset voltage is

$$\Delta V_{oo} \cong R_F \Delta I_{io} \cong 1\,\text{M}\Omega(15\,\text{nA}) = 15\,\text{mV}$$

Thus as the temperature increases from 20°C to 60°C, the output offset voltage increases from its initial 0 V to 15 mV.

Sometimes the drift in input bias current due to temperature changes is specified in terms of picoamperes per degree Celsius over a specified temperature range. For example, some varactor-input Op Amps have a specified drift of 0.001 pA/°C over the temperature range from +10°C to 70°C. Some FET-input Op Amp's data sheets might show that current simply doubles for every 10° rise in temperature in the range from −25°C to +85°C.

Needless to say, specifications on drift, whether on curves or in tables, enable us to predict the maximum drift we might have in the output voltage level under known temperature variations.

7-3   Other Temperature-Sensitive Parameters

A drift in the Op Amp's output voltage level is not the only possible result of temperature changes. Figure 7-6 shows a typical collection of other temperature-sensitive characteristics. The curves in Fig. 7-6a show the drift of three parameters with temperature changes on a relative scale. In this case, relative scale indicates multiplying factors that enable us to determine the changes in the three parameters if the temperature drifts away from 25°C. For example, Fig. 7-6a shows that the transient response at 70°C is 1.1 times the transient response at 25°C. On this same graph, the closed-loop BW at 70°C is shown to be 0.9 of its value at 25°C. The slew rate is shown to be quite constant, especially with temperatures above 25°C.

The transient response of an Op Amp is the time (usually in microseconds) required for its output voltage to rise from 10% to 90% of its final value under *small-signal* conditions. A typical transient response waveform is shown in Fig. 7-7. Note that the voltage level here is much smaller than that used to define the slew rate.

As shown in Fig. 7-6b, the input resistance R_i of the typical Op Amp increases with an increasing temperature. This could be a problem if we drive a noninverting amplifier with a high internal resistance signal source. A drifting input resistance, which is the load resistance on the signal source, can cause a drifting input signal amplitude.

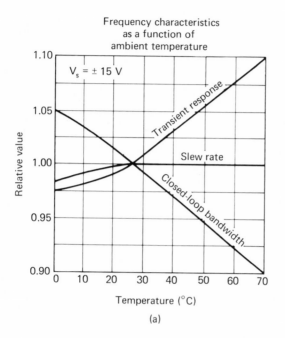

(a)

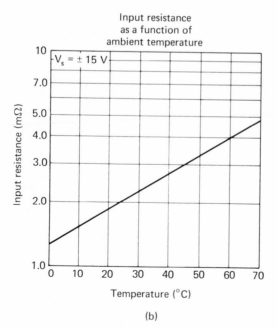

(b)

Fig. 7-6 Temperature-sensitive parameters of the typical Op Amp.

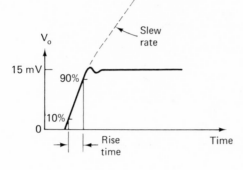

Fig. 7-7 Typical transient response waveform. The rise time is the time the output voltage can change from 10% to 90% of its final level under small-signal conditions.

7-4   Programmable Op Amps

Some manufacturers are marketing Op Amps whose characteristics can be varied somewhat to suit special applications. These are called programmable Op Amps and they have an additional input *bias* terminal which is referred to as the I_{SET} input as shown in Fig. 7-8. By controlling the current I_{SET} in this terminal by external means, we can adjust some of the Op Amp's parameters to values that will optimize its performance in any given application. The bias current I_{SET} can be controlled or selected with an externally wired resistor R_{SET} that can be connected to ground or to one of the supply voltages as shown in Fig. 7-9. For the circuit in Fig. 7-9a,

$$I_{SET} = \frac{V - 0.7}{R_{SET}} \tag{7-6}$$

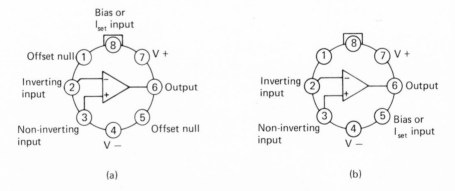

(a)   (b)

Fig. 7-8 Typical programmable Op Amp packages: (a) type 776 (Fairchild Semiconductor); (b) type 3080 (RCA).

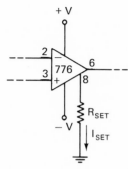

(a)

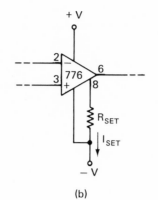

(b)

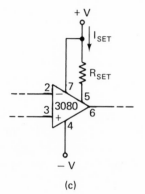

(c)

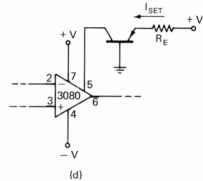

(d)

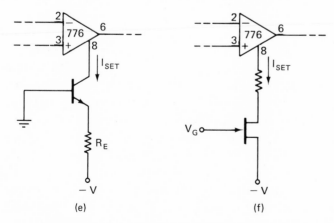

(e)   (f)

Fig. 7-9 Typical methods of establishing set current I_{SET} in programmable Op Amps.

where V is the dc voltage at the $+V$ terminal with respect to ground. For the circuit in Fig. 7-9b,

$$I_{SET} = \frac{V' - 0.7\ V}{R_{SET}} \qquad (7\text{-}7)$$

where V' is the voltage across the $+V$ and $-V$ terminals and is the sum $|+V| + |-V|$.

For the circuits in Fig. 7-9d and Fig. 7-9e,

$$I_{SET} = \frac{V - 0.7}{R_E} \qquad (7\text{-}8)$$

where V is the voltage to which the resistor R_E is connected.

The recommended range of current I_{SET} and how it is to be established or controlled varies in programmable Op Amps of different types and with different manufacturers. The manufacturers' data sheets, however, provide all the specific information needed to use their products. For example, typical programmable Op Amp characteristics are shown in Fig. 7-10. The curve in Fig. 7-10a shows how the input bias current I_B varies with current I_{SET}. Obviously, the input bias current I_B is less ideal (larger) with larger values of I_{SET}. Similarly then the drift of this input bias current I_B with temperature changes is undesirably larger with a larger I_{SET}, as shown by the curve in Fig. 7-10b. Related to I_B drift is the drift in input offset current I_{io} which also is larger, and therefore potentially more troublesome, with larger values of I_{SET}. See the curve in Fig. 7-10d. Curve c of the figure shows that the drift in the input offset voltage V_{io} is adjustable with I_{SET} and that this drift can be reduced to almost zero if $I_{SET} = 1\ \mu A$. All of these curves present a good case for selecting a small I_{SET}. However, note in the curve of Fig. 7-10h that the slew rate is faster (more ideal) with larger values of I_{SET}. Similarly, we see that the output voltage swing capability and the $CMRR$ improve with larger values of I_{SET} according to the graphs in Fig. 7-10f, g, and i. With this programmable Op Amp then, we can use more or less I_{SET}, depending on what the application demands. If very low drift is most essential, we will use a relatively small I_{SET}. If, on the other hand, a large output swing capability or high $CMRR$ are premium, we will use larger I_{SET} values.

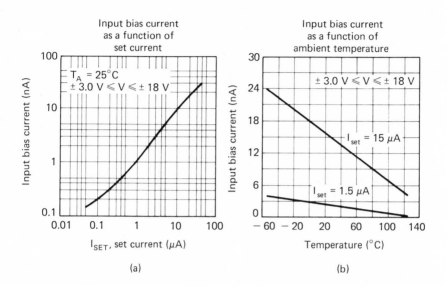

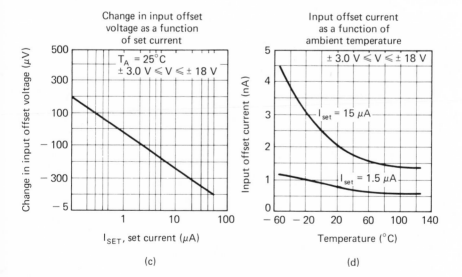

Fig. 7-10 Typical characteristics of the programmable type 776 Op Amp; voltage V is the dc supply voltage. (Courtesy of Fairchild Semiconductor Components Group, Camera and Instrument Corp.)

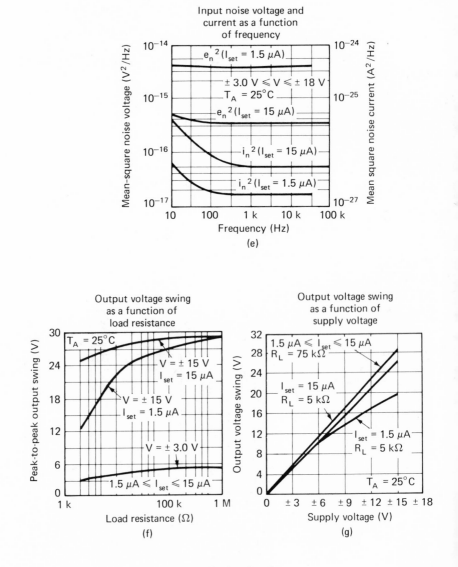

Fig. 7-10 *(continued)*

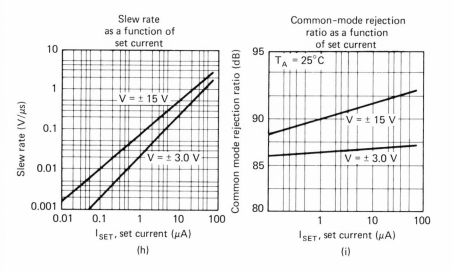

Fig. 7-10   *(continued)*

7-5   Varactor- and Chopper-stabilized Op Amps

The monolithic IC Op Amps, the main theme of this text, have attractive features such as low cost and small size. However, where *extremely* low drift and bias currents are required for some stringent application, certain types of hybrid Op Amps are used. Varactor-diode-input and chopper-stabilized Op Amps are in this family of hybrids. The varactor-input Op Amps, in particular, offer the designer low bias currents, on the order of 0.01 pA, and low bias-current *drifts*, about 0.001 pA/°C. They also feature extremely large input resistance: approximately $10^{14}\,\Omega$ looking into the noninverting input to ground, as shown in the equivalent circuit of Fig. 7-11. Chopper-stabilized Op Amps also have low drift, on the order of 0.5 pA/°C, compared to typical monolithic types.

Detailed discussions of how varactor-input and chopper-stabilized Op Amps work internally are beyond the scope of this applications-oriented book. Generally, they work on the principle of modulating or chopping low-frequency and dc input signals V_s. This in effect changes them to a higher ac frequency whose amplitude is proportional to the amplitude of the applied signals V_s. The modulated or chopped signal is processed through

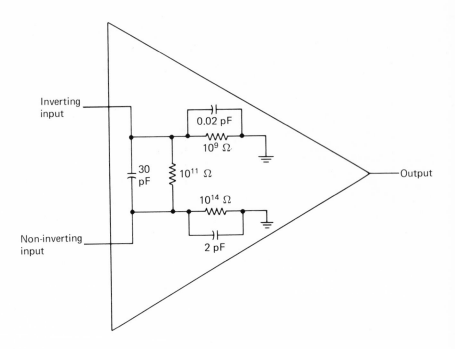

Fig. 7-11 Equivalent circuit of the input of a varactor-type Op Amp.

an ac amplifier and is then demodulated or filtered. See Fig. 7-12. The demodulator or low-pass filter restores the original applied waveform of V_s, which is then applied to a main dc-type amplifier. Higher frequencies of input signals V_s are directly applied to the input of the main dc amplifier through a high-pass filter and are not processed through the modulator or chopper channel. The drift of I_B, I_{io}, and V_{io} in this main dc amplifier appear *smaller* at the input terminals of the overall amplifier by the gain factor of the ac stage.

7-6   Channel Separation

If a monolithic Op Amp package contains two amplifiers (dual Op Amp), a parameter called *channel separation* is specified in its data sheets. It tells us how much interaction we can expect between the two amplifiers. If a signal is applied to the input of one amplifier in a dual package, some signal will appear at the output of the other Op Amp even though it has no input

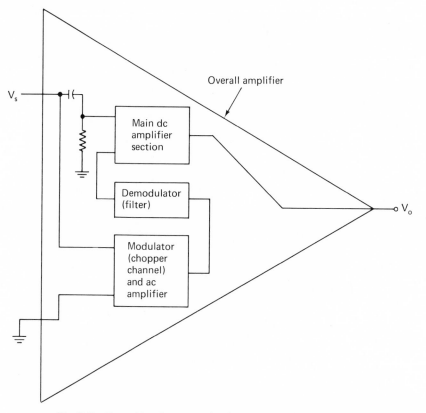

Fig. 7-12   Dc and low-frequency signals are processed by a modulator or a chopper-channel section.

signal applied. This interaction exists because of the close physical proximity, which causes electrical coupling, between the two Op Amps built on a single semiconductor chip.

The 747, whose packages are shown in Fig. 7-13, contains essentially two 741 Op Amps. Its channel separation is typically specified at a minimum of 100 dB. This means that if one of the Op Amps is driven while the other is not, the output signal of the undriven one will be at least 100 dB below the signal output of the driven amplifier.

EXAMPLE 7-3

If one of the Op Amps of a 747 is driven so that its output signal is 15-V peak to peak at 400 Hz, how much 400-Hz signal might be at the output of the other amplifier even though it is not driven.

Dual-in-line packages

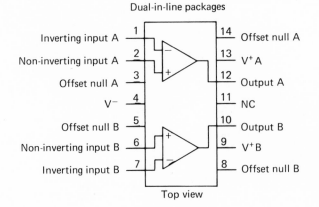

Top view

Metal can package

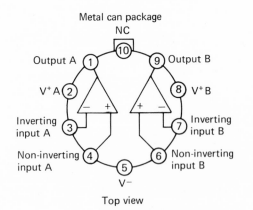

Top view

Flat package

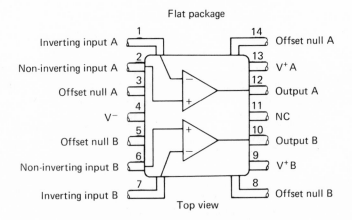

Top view

Fig. 7-13 Packages containing dual Op Amps.

Answer. Since the 747's channel separation is at least 100 dB, which is equivalent to a ratio of 10^5, the output of the undriven amplifier will not be more than

$$\frac{15\,\text{V}}{10^5} = 150\,\mu\text{V peak to peak}$$

7-7   Cleaning PC Boards and Guarding Input Terminals

As mentioned previously, some applications require Op Amps with low bias currents. Certain types of high-beta transistor-input Op Amps, FET-input Op Amps, varactor-input Op Amps, and chopper-stabilized Op Amps are made for such applications. Use of such low-bias-current Op Amps presents new problems, however. When the low-bias-current Op Amp is used on a printed circuit (PC) board, the circuit tends to behave erratically and may drift with temperature changes, especially in higher temperature ranges, if certain precautions are not taken.

A drift problem can be caused by PC board leakage currents. These currents can easily exceed the bias currents of the Op Amp, especially if it is a low-bias-current type. To reduce the possibility of such drift, the printed circuit board must be thoroughly cleaned with alcohol or trichlorethylene to remove all solder flux. The board can then be coated with silicone rubber or epoxy to keep its surface clean.

Even a clean PC board has finite resistance. At 125°C, the resistance between parallel conductors 1 inch long with 0.05-inch separation is about $10^{11}\,\Omega$. With just a few volts across such conductors, the leakage current between them will easily run up to hundreds of picoamperes, which is much larger than bias currents in some high-performance Op Amps. Therefore, because of the close spacing between pins, the dc supply voltages can force currents through the board and Op Amp's inputs. Since such currents might be large relative to the Op Amp's bias current, they can drive the output into saturation or at least cause a drifting output voltage. This problem is reduced by surrounding the inputs with a conductive *guard*, which is terminated to some low-impedance point that is essentially at the same potential as the inputs. This effectively prevents current flow through the board material into the Op Amp's inputs.

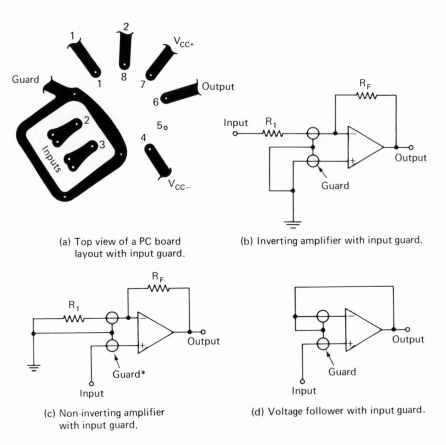

(a) Top view of a PC board layout with input guard.

(b) Inverting amplifier with input guard.

(c) Non-inverting amplifier with input guard.

(d) Voltage follower with input guard.

Fig. 7-14

Figure 7-14 shows a PC board layout with a guard around the inputs, along with its schematic representation and proper termination in popular circuit types.

7-8　Protecting Techniques

It takes on the order of just a few millivolts across an Op Amp's inverting and noninverting inputs to drive its output into saturation. Large differential inputs can ruin it. If large input peaks or transients are expected across the inputs, the Op Amp can be protected as shown in Fig. 7-15. The

* This guard should be tied to a dc voltage equal to the dc average input signal voltage.

diodes remain nonconducting and therefore do not affect the input signals as long as these signals are small as they normally should be. Large input signals, however, drive the diodes into conduction. Thus the differential input voltages are limited to a few hundred millivolts—the forward voltage drop of each diode.

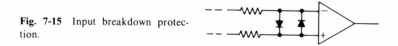

Fig. 7-15 Input breakdown protection.

Most IC Op Amps have built-in *output short-circuit protection*. Note that this quality is specified in Appendix SI. Some types, such as the 702 and the 709, can tolerate an output short circuit for just a short time and can deliver about 75 mA. If we keep such current drain for too long, the unprotected Op Amp will be ruined. A resistor in series with the output, as in Fig. 7-16, will keep the current drain within safe limits. For the 709, the manufacturers recommend a 200-Ω series output resistor if the output may be short-circuited to ground. Of course, if a fixed load resistor greater than 200 Ω is used, there is no need for using a series output resistor even though the Op Amp is not internally protected.

Fig. 7-16 Output short-circuit protection.

Op Amps can be ruined if the dc supply voltages are connected in the wrong polarities. Diodes in series with the dc supply leads, such as in Fig. 7-17, will protect the Op Amp from such an accident.

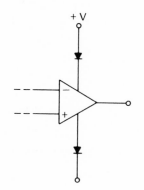

Fig. 7-17 Dc supply voltage reversal protection.

If unregulated or poorly regulated dc supplies are used, the Op Amp can be protected from excessive supply voltage with a zener (regulator diode) as shown in Fig. 7-18. Thus if the Op Amp's maximum dc supply voltages are specified at ± 17 V, a zener with a breakdown voltage of 34 V or less would be used.

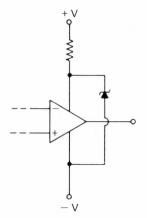

Fig. 7-18 Dc supply overvoltage protection. ·

Some types of Op Amps are subject to *latch-up* if driven into saturation. Latch-up means that in Op Amp's output stays saturated even though the input signal that originally caused the saturation has been removed. The type 709 Op Amp can latch up, but an externally wired diode, as shown in Fig. 7-19, will help prevent it. Some IC Op Amps are specified by their manufacturers as having no latch-up problems and do not require externally wired protection.

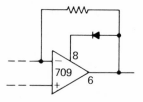

Fig. 7-19 Protection against latch-up on the type 709 Op Amp.

REVIEW QUESTIONS

1. If a hypothetical Op Amp has an infinite power supply rejection ratio, how stringent should its power supplies' regulation and filtering be?

2. What are some causes of drift in the output voltage level of an Op Amp with temperature changes?

3. From which would we expect more drift; an Op Amp with relatively large or a relatively low input bias current? Why?

4. In the circuit of Fig. 7-4a, if given the change in the input bias current and the change in the input offset current with temperature, which of these specifications enables us to predict the output voltage drift? Why?

5. In the circuit of Fig. 7-5a, if we know the Op Amp's change in input bias current and its change in input offset current with temperature, which of these specifications enables us to predict the output voltage drift? Why?

6. Name five Op Amp parameters that drift with temperature changes

7. What outstanding features do the varactor-input and the chopper-stabilized Op Amps have compared to the general-purpose IC Op Amps?

8. What is a programmable Op Amp?

9. What PC board precautions should be followed when Op Amps having extremely low bias currents are used?

10. If large output signal swing is expected from an Op Amp, should it be selected from types specifying "No Latch-Up"? Why?

11. If the ambient temperature is subject to considerable changes, which is more stable, with a given type of general-purpose Op Amp, the circuit in Fig. 7-4a or the circuit in Fig. 7-5a? Why?

12. What purpose does the resistor R_2 serve in the circuit of Fig. 7-5c?

PROBLEMS

1. In the circuit of Fig. 7-2, suppose that we null the output when the voltage across the $+V$ and $-V$ supply terminals measures 24 V. This voltage then drifts from 20 V to 28 V. If the Op Amp's $PSRR = 100\,dB$, what changes can we expect in its (a) input offset voltage, and (b) output voltage level?

2. In the circuit of Fig. 7-5b, what drift in input offset voltage and output voltage can we expect if $R_1 = R_a = 10\,k\Omega$, $R_F = R_b = 1\,M\Omega$, the power supply sensitivity $S = 100\,\mu V/V$, and the voltage across the $+V$ and $-V$ supply terminals drifts in the range from 20 V to 25 V?

3. In the circuit described in Problem 1, how much 60-Hz hum can we expect across the load if there is a 4-mV rms, 60-Hz voltage across the $+V$ and $-V$ pins?

4. In the circuit described in Problem 2, how much 60-Hz hum can we expect across the load if a 6-mV rms, 60-Hz voltage is measured across the $+V$ and $-V$ terminals?

5. In the circuit of Fig. 7-4a, suppose that $R_s = 1000\,\Omega$, $R_1 = 20\,\mathrm{k\Omega}$, $R_F = 10\,\mathrm{M\Omega}$, and the circuit is nulled at 25°C. What output offset could we expect at 65°C if the Op Amp is an FET type whose bias current is 30 pA at 25°C and whose bias current doubles for every 10°C rise in temperature?

6. In the circuit of Fig. 7-4a, suppose that $R_s = 100\,\Omega$, $R_1 = 1\,\mathrm{k\Omega}$, $R_F = 20\,\mathrm{k\Omega}$, and the circuit is nulled at 20°C. What output offset voltage can we expect at 0°C if the Op Amp has the characteristics shown in Fig. 7-3?

7. If in the circuit of Fig. 7-5a, $R_s = 200\,\Omega$, $R_1 = 2\,\mathrm{k\Omega}$, $R_2 = 2\,\mathrm{k\Omega}$, $R_F = 20\,\mathrm{k\Omega}$, and the output offset is 500 mV at 20°C, what is the approximate maximum possible output offset voltage at (a) 0°C and at (b) 60°C? Assume that the Op Amp has characteristics as shown in Fig. 7-3.

8. Assume that in the circuit of Fig. 7-4b, $R_1 = 1\,\mathrm{k\Omega}$, $R_2 = 1\,\mathrm{k\Omega}$, $R_s = 100\,\Omega$, and $R_F = 10\,\mathrm{k\Omega}$. If the Op Amp has the characteristics shown in Fig. 7-3, and if the circuit is nulled at 20°C, what output offset can we expect at temperatures (a) $-20°C$ and (b) 100°C?

9. If an Op Amp has the characteristics shown in Fig. 7-6a and has a 100-ns transient response at 25°C, what is its transient response at (a) 0°C and at (b) 70°C?

10. If an Op Amp has the characteristics shown in Fig. 7-6a and has a closed-loop bandwidth of 40,000 Hz at 25°C, what are its bandwidths at temperatures (a) 0°C and (b) 70°C?

CHAPTER 8

LINEAR
APPLICATIONS
OF OP AMPS

A linear circuit generally is one whose output is an amplified version of the input signal or a predictable function of the input signal. All circuits discussed in previous chapters are linear types. A collection of a few common ones is given in Fig. 8-1. Other linear circuits discussed in this chapter are ac amplifiers, summing circuits, integrators, differentiators, regulators, and filters.

All linear circuits behave linearly and predictably over a limited range. The maximum positive and negative output voltages of the Op Amp are the limits within which it will be a linear device. There are some nonlinear applications of Op Amps, particularly in digital and switching circuits, in which their outputs are purposely driven into positive or negative saturation, but these are discussed in later chapters.

8-1   Op Amps as ac Amplifiers

The Op Amp applications discussed till now were mainly dc amplifiers. In other words, their output voltages change in respose to changes in dc input levels. All the circuits in Fig. 8-1 were discussed previously and are dc type amplifiers. Of course, they respond to ac input signals too, provided the frequencies are not too high. In some applications, the circuit designer needs the ac response characteristics of the Op Amp but does not need or want its dc response capability. An audio signal amplifying system could be an example. Even a small dc input offset voltage in the first stage can easily be amplified to a value large enough to saturate the following Op Amp stages if the stages are directly coupled. Capacitive coupling, as shown in Fig. 8-2,

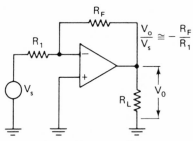

(a) Basic inverting amplifier.

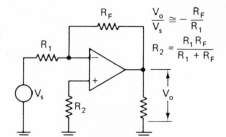

(b) Inverting amplifier with improved stability.

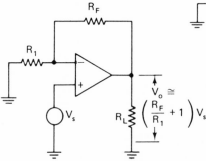

(c) Basic non-inverting amplifier.

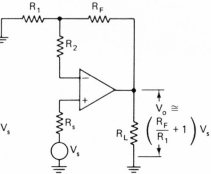

(d) Non-inverting amplifier with improved stability.

$$R_2 = R_s - \frac{R_1 R_F}{R_1 + R_F} \quad \text{if} \quad R_s > \frac{R_1 R_F}{R_1 + R_F}$$

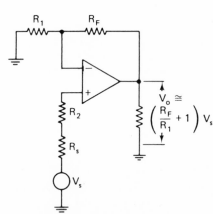

(e) Non-inverting amplifier with improved stability.

$$R_2 = \frac{R_1 R_F}{R_1 + R_F} - R_s$$

$$\text{if} \quad R_s < \frac{R_1 R_F}{R_1 + R_F}$$

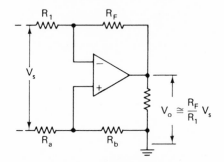

(f) Basic differential input amplifier.

$R_1 = R_a$ and $R_F = R_b$
for good CMRR

Fig. 8-1 Common linear Op Amp applications.

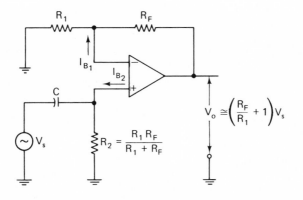

Fig. 8-2   An ac noninverting amplifier; I_{B_1} and I_{B_2} are dc bias currents.

between stages is a simple way of eliminating dc level amplification from stage to stage. For example, the signal source V_s in Fig. 8-2 could be a preceding amplifier with a dc component in its output, and the coupling capacitor C effectively blocks the dc component, thus preventing it from affecting the next stage. The resistor R_2 must be used because it provides a dc path between the noninverting input and ground. Without R_2, the coupling capacitor C would become charged by the bias current I_{B_2} and would cause a dc input voltage on the noninverting input and a resulting output offset.

Since the circuit in Fig. 8-2 is basically a noninverting amplifier, its closed-loop gain is

$$A_v \cong \frac{R_F}{R_1} + 1 \tag{3-6}$$

at frequencies within its bandwidth. The low-frequency limit of the bandwidth is determined largely by the input resistance and the size of the coupling capacitor C.*

An inverting ac amplifier is shown in Fig. 8-3. If the reactance of C is negligible, its gain is

$$A_v \cong -\frac{R_F}{R_1} \tag{3-4}$$

at frequencies within its bandwidth. If high gain or a large feedback resistor R_F is used, a resistor equal to R_F can be placed between the noninverting input and ground to reduce the output offset and drift caused by bias current.

* See Appendix A3 for a method of selecting the proper coupling capacitor values.

With negligible reactances of the coupling capacitors C, the signal source V_s sees R_2 as the stage input impedance in the circuit of Fig. 8-2, and it sees R_1 as the input resistance in the circuit of Fig. 8-3. These impedances are relatively low and in cases where the ac signal source V_s must see a very large stage input impedance, the amplifier's input impedance can be *bootstrapped* as shown in Fig. 8-4. In this circuit, which is a voltage follower, the output signal V_o is applied to the bottom of R_1 via capacitor C_2. Simultaneously the input signal V_s is applied to the top of R_1 through the coupling capacitor C_1. Therefore, the signal potential difference across R_1 is $V_s - V_o$. But since $V_s \cong V_o$ in this voltage follower, their difference is extremely small, which means that the signal current through R_1 is also extremely small. Since the signal current in R_1 is also the current drain from the signal source V_s, the impedance seen by the signal source V_s is extremely large.

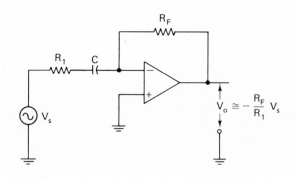

Fig. 8-3   Ac inverting amplifier.

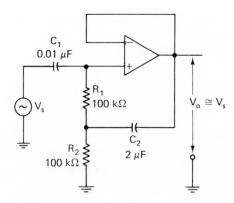

Fig. 8-4 Ac voltage follower with input impedance bootstrapped (increased).

8-2 Op Amp ac Amplifier with a Single Power Supply

If coupling capacitors are used on the input and output of an Op Amp circuit, a nulled output is not essential. Since the output capacitor blocks dc, an output offset voltage or a dc component in the output signal has little effect on the circuit's operation, especially with small signals. This eliminates the need for two dc supply voltages: one positive and one negative with respect to ground. One supply with a *split zener* arrangement can be used as shown in Fig. 8-5 with little or no loss of output signal peak-to-peak capability. The zeners are selected so that the sum of their drops is a little less than the minimum value of the supply voltage $+V$. For example, if the minimum supply voltage is 15 V, two 6-V zeners can be used with the remaining 3 V dropped across the resistor R. The 6-V drop across D_2 is applied to the noninverting input, and it lifts the voltage at the output pin 6 to about 6 V. That is, dc into the noninverting input effectively sees the circuit as a voltage follower because the capacitor C_1 blocks dc current flow through and prevents dc voltage drops across R_F and R_1. Essentially, then, all of the dc at the output pin 6 is fed back to the inverting input pin 2, resulting in unity dc gain. With ac values of V_s, however, the gain $A_v = -R_F/R_1$ at frequencies that cause the reactance of C_1 to be negligible.

The resistor R in the circuit of Fig. 8-5 is selected to drop the difference in the supply voltage $+V$ and the total drop across both zeners, and it must be capable of conducting a few milliamperes more than the maximum dc current expected to be drawn by the Op Amp. Resistor R_2 is selected to be equal to R_F to minimize the dc differential input voltage, which assures us of good peak-to-peak output signal capability. C_3 and C_4 are decouplers which keep signal out of the zeners and thus prevent signal from modulating the power supply voltage.

EXAMPLE 8-1

In the circuit of Fig. 8-5, if the reactances of all capacitors are negligible, $R_1 = 1 \text{ k}\Omega$, $R_2 = 20 \text{ k}\Omega$, $R_F = 20 \text{ k}\Omega$, $R = 80 \Omega$, $+V = 24 \text{ V}$, each zener drop is 10 V, and the input signal V_s is sinusoidal with a 0.2 V peak, what are the instantaneous peak positive and peak negative voltages at pin 6 and at the right side of C_2? What current is in resistor R?

Answer. With 10 V across each zener, the 10 V of D_2 is applied to the noninverting input, causing very nearly 10 V dc at the output pin 6.

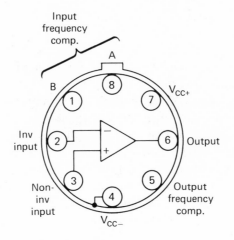

(a)

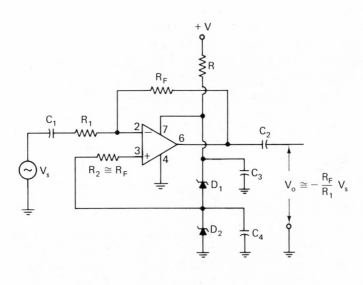

(b)

Fig. 8-5 (a) 709 in plug-in package (top view); (b) 709 wired to work as an ac amplifier off a single power supply.

With ac signals, the circuit gain is

$$A_v \cong -\frac{R_F}{R_1} = -\frac{20\text{ k}\Omega}{1\text{ k}\Omega} = -20 \qquad (3\text{-}4)$$

The 0.2-V peak ac input therefore is amplified to $-20(0.2\text{ V}) = -4\text{ V}$ peak. The negative sign simply means that the input and output waveforms are out of phase. This 4-V peak output rides on the 10-V dc component at pin 6, causing the voltage at this point to swing to 14-V peaks on positive alternations and to 6-V peaks on negative alternations. On the right side of C_2 the dc component is gone, and the voltage swings to 4 V on positive alternations and to -4 V on negative alternations.

The zeners drop a total of 20 V, leaving the remaining 4 V across the resistor R. By Ohm's law, the dc current through it is $4\text{ V}/80\,\Omega = 50\text{ mA}$.

8-3   Summing and Averaging Circuits

The circuit in Fig. 8-6 is the basic inverting summing and averaging circuit. The value and polarity of the output voltage V_o is determined by the sum of the input voltages V_1 through V_n and by the values of the externally wired resistors. Since the noninverting input is at ground potential, the inverting input and the right sides of input resistors R_a through R_n are

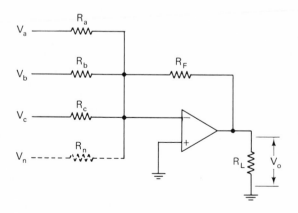

Fig. 8-6   Inverting basic summing or averaging circuit.

virtually grounded too. Thus the currents in the input resistors can be shown, by Ohm's law, as

$$I_a \cong \frac{V_a}{R_a}$$

$$I_b \cong \frac{V_b}{R_b}$$

$$I_c \cong \frac{V_c}{R_c}$$

$$I_n \cong \frac{V_n}{R_n}$$

Similarly, the current in R_F is

$$I_F \cong \frac{-V_o}{R_F}$$

where the negative sign indicates that V_o is out of phase with the net voltage at the inverting input. Since the sum of the input currents is equal to the feedback current, assuming that the Op Amp is ideal, we can show that

$$I_F \cong I_a + I_b + I_c + \cdots + I_n$$

or

$$\frac{-V_o}{R_F} \cong \frac{V_a}{R_a} + \frac{V_b}{R_b} + \frac{V_c}{R_c} + \cdots + \frac{V_n}{R_n}$$

Multiplying both sides of the above equations by R_F shows that, generally, the output voltage is

$$V_o \cong -R_F\left(\frac{V_a}{R_a} + \frac{V_b}{R_b} + \frac{V_c}{R_c} + \cdots + \frac{V_n}{R_n}\right) \tag{8-1}$$

If all input resistors are equal, say

$$R_a = R_b = R_c = R_n = R$$

Eq. (8-1) simplifies to

$$V_o \cong -\frac{R_F}{R}(V_a + V_b + V_c + \cdots + V_n) \tag{8-2}$$

If $R_F = R$, this equation simplifies further to

$$V_o \cong -(V_a + V_b + V_c + \cdots + V_n) \tag{8-3}$$

This last equation shows that the circuit in Fig. 8-6 can be used to find the negative sum of any number of input voltages.

This circuit can also average the input voltages by using a ratio of R_F/R that is equal to the reciprocal of the number of voltages being averaged. The ratio R_F/R is selected so that the sum of the input voltages is divided by the number of input voltages applied.

Of course, the above equations apply assuming that the Op Amp is properly nulled, that is, $V_o = 0$ V if all inputs are 0 V. The effect that input bias current has on the output offset and stability can be reduced if we use a resistor R_2 between the noninverting input and ground, as shown in Fig. 8-7,

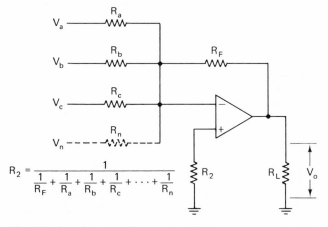

Fig. 8-7 Inverting summing or averaging circuit with improved stability.

where its value is

$$R_2 = \frac{1}{1/R_F + 1/R_a + 1/R_b + 1/R_c + \cdots + 1/R_n} \tag{8-4}$$

The input voltage sources and resistors can be connected to the non-inverting input of an Op Amp as shown in Fig. 8-8a. These sources and resistors can be replaced with a Norton's equivalent circuit, as shown in Fig. 8-8b, where

$$I_{eq} = \frac{V_a}{R_a} + \frac{V_b}{R_b} + \frac{V_c}{R_c} + \cdots + \frac{V_n}{R_n} \tag{8-5}$$

and

$$R_{eq} = \frac{1}{1/R_a + 1/R_b + 1/R_c + \cdots + 1/R_n} \tag{8-6}$$

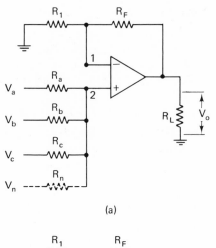

(a)

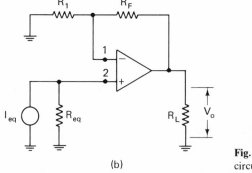

(b)

Fig. 8-8 Noninverting averaging circuit.

Since the resistance looking into the noninverting input is very large, practically all of the current I_{eq} is forced through R_{eq}, resulting in a voltage at the noninverting input 2, that is

$$V_2 = R_{eq}I_{eq} \qquad (8\text{-}7)$$

If all of the input resistors are equal, voltage V_2 is the average of the input voltages. The resulting output voltage is

$$V_o \cong \left(\frac{R_F}{R_1} + 1\right)V_2 \qquad (8\text{-}8)$$

EXAMPLE 8-2

Determine the value and polarity of the output voltage in each of the circuits in Fig. 8-9.

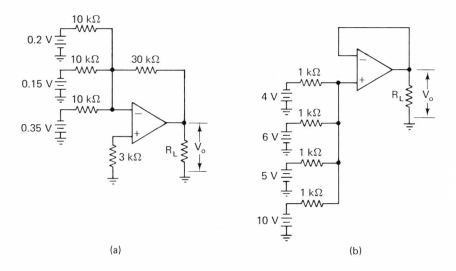

Fig. 8-9

Answer. The circuit shown in Fig. 8-9a is an inverting summing amplifier. Since all of its input resistors are equal, the output voltage can be determined with Eq. (8-2), that is,

$$V_o \cong -\frac{30\,\text{k}\Omega}{10\,\text{k}\Omega}(0.2\,\text{V} - 0.15\,\text{V} - 0.35\,\text{V})$$

$$= -3(-0.3\,\text{V}) = +0.9\,\text{V}$$

The circuit in Fig. 8-9b is a noninverting type whose input voltages and resistances are replaced with a Norton's equivalent circuit as shown, where

$$I_{eq} = \frac{-4\,\text{V}}{1\,\text{k}\Omega} + \frac{6\,\text{V}}{1\,\text{k}\Omega} + \frac{5\,\text{V}}{1\,\text{k}\Omega} + \frac{-10\,\text{V}}{1\,\text{k}\Omega} \qquad (8\text{-}5)$$

$$= -4\,\text{mA} + 6\,\text{mA} + 5\,\text{mA} - 10\,\text{mA} = -3\,\text{mA}$$

and

$$R_{eq} = \frac{1\,\text{k}\Omega}{4} = 250\,\Omega \qquad (8\text{-}6)$$

Therefore, the voltage at the noninverting input

$$V_2 = 250\,\Omega(-3\,\text{mA}) = -0.75\,\text{V} \qquad (8\text{-}7)$$

which is the average of the input voltages. We can recognize that this Op Amp is connected to work as a voltage follower, which means that the output voltage V_o is the same as the input voltage or -0.75 V in this case. We could have used Eq. (8-8) and obtained the same result.

8-4   The Op Amp As an Integrator

The Op Amp in Fig. 8-10 is connected to work as an integrator. Its output voltage waveform is the negative integral of the input voltage waveform for properly selected values of R and C. Its operation can be analyzed as follows: Assuming that the Op Amp is ideal, the inverting input 1 and the right side of the input resistor R are at ground potential because the noninverting input is grounded. Therefore, the applied voltage V_s appears across R and the current in this resistor is

$$I = \frac{V_s}{R} \qquad (8\text{-}9)$$

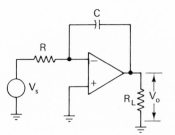

Fig. 8-10   Op Amp wired to work as an integrator.

Because of the very large resistance looking into the inverting input, practically all of this current is forced through the capacitor C, which changes the voltage across it. Generally, the current through and the voltage across a capacitor are related by the following equations:

$$i_c = C\frac{dv_c}{dt} \qquad (8\text{-}10\text{a})$$

$$v_c = \frac{1}{C}\int i_c\,dt \qquad (8\text{-}10\text{b})$$

Since the left side of the capacitor C in Fig. 8-10 is virtually grounded, then the output voltage V_o is the voltage across the capacitor, and therefore Eq. (8-10b) can be modified to

$$V_o = -\frac{1}{C}\int I\,dt$$

where I is the current being forced through the capacitor. Substituting the right side of Eq. (8-9) into the above results in

$$V_o = -\frac{1}{C}\int \frac{V_s}{R}\,dt$$

or

$$V_o = -\frac{1}{RC}\int V_s\,dt \qquad (8\text{-}11)$$

where the negative sign represents the phase-inverting property of this circuit. If we select the R and C values so that their product is 1, the above equation simplifies to

$$V_o = -\int V_s\,dt \qquad (8\text{-}12)$$

EXAMPLE 8-3

If in the circuit of Fig. 8-10, $R = 1\,M\Omega$, $C = 1\,\mu F$, and V_s has the waveform in Fig. 8-11a, sketch the resulting waveform of the output V_o on the scale in Fig. 8-11b. Assume that the Op Amp is ideal and that the capacitor C is initially uncharged.

Answer. The step voltage function applied can be expressed as a constant function beginning at $t = 0.1$ s. Since the RC product is 1, Eq. (8-12) applies, and we can show that

$$V_o = -\int V_s\,dt$$

$$= -\int 8\,dt$$

$$= -8t + V_i = -8t$$

where V_i represents the initial voltage on the capacitor, which in this case is 0 V. Thus, with a constant voltage function applied to the input of the integrator, the output is a ramp function with a negative slope.

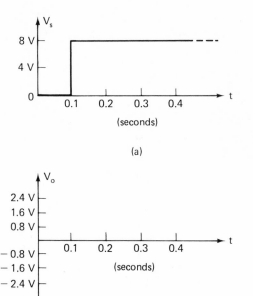

(a)

(b)   **Fig. 8-11**

Thus the output voltage is a linear function of time, and we can show that since

$$V_{o(t)} = -8t$$

then

$$V_{o(0.1s)} = -8 \times 0.1 = -0.8 \text{ V}$$

and then

$$V_{o(0.2s)} = -8 \times 0.2 = -1.6 \text{ V}$$

etc. Sketching these versus time results in an output waveform as shown in Fig. 8-12. A number of other input and output voltage waveforms of this circuit are shown in Fig. 8-13.

As with the previously discussed Op Amp applications, the integrating circuit of Fig. 8-10 can be made less prone to drift and have less output offset if the proper resistance value is placed between the noninverting input and ground. In this case, this resistance is simply made equal to the input resistance R.

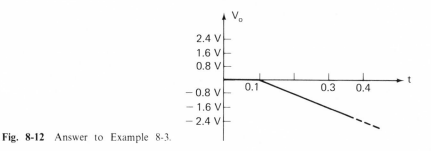

Fig. 8-12 Answer to Example 8-3.

8-5 The Op Amp As a Differentiator

The Op Amp can be wired to work as a differentiator. As such, its output is essentially the derivative of the input voltage waveform. The basic differentiator is the circuit shown in Fig. 8-14a. The right side of the input capacitor C is virtually grounded, and therefore the voltage across it is the input voltage V_s. Current I flows to charge or discharge C only when the input voltage changes. Thus, by substituting V_s for v_c and I for i_c in Eq. (8-10a), we can show that

$$I = C\frac{dV_s}{dt} \qquad (8\text{-}13)$$

Since the resistance looking into the inverting input is very large, any existing input current I is forced up through the feedback resistor R_F. But the left side of R_F is virtually grounded, and therefore the voltage drop across it is the output voltage V_o. By Ohm's law then

$$V_o = R_F I \qquad (8\text{-}14)$$

Rearranging and substituting this last equation into Eq. (8-13), we can show that

$$\frac{V_o}{R_f} = -C\frac{dV_s}{dt}$$

or

$$V_o = -R_F C\frac{d}{dt}(V_s) \qquad (8\text{-}15)$$

Apparently, the output voltage V_o is the derivative of the input voltage V_s times the negative product of R_F and C.

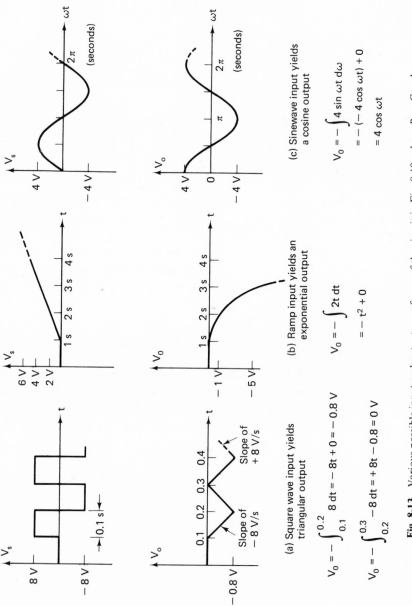

(a) Square wave input yields triangular output

$$V_0 = -\int_{0.1}^{0.2} 8\,dt = -8t + 0 = -0.8\text{ V}$$

$$V_0 = -\int_{0.2}^{0.3} -8\,dt = +8t - 0.8 = 0\text{ V}$$

(b) Ramp input yields an exponential output

$$V_0 = -\int 2t\,dt$$

$$= -t^2 + 0$$

(c) Sinewave input yields a cosine output

$$V_0 = -\int 4\sin\omega t\,d\omega$$

$$= -(-4\cos\omega t) + 0$$

$$= 4\cos\omega t$$

Fig. 8-13 Various possible input and output waveforms of the circuit in Fig. 8-10, where $R \times C = 1$.

Unfortunately, the circuit in Fig. 8-14 has some practical problems. Because the ratio of the feedback resistance, R_F, to the input capacitor's reactance, X_C, rises with higher frequencies, this circuit's gain increases with frequency. This tends to amplify the high-frequency noise generated in the system, and the resulting output noise can completely override the differentiated signal.

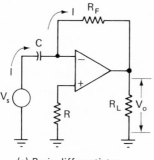

(a) Basic differentiator

$$V_o = -R_F C \frac{d}{dt} V_s$$

$$R = R_F$$

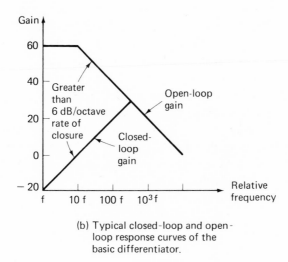

(b) Typical closed-loop and open-
loop response curves of the
basic differentiator.

Fig. 8-14

Another problem with this basic differentiator is its tendency to be unstable. In Chapter 6 we learned that, if the closed-loop and open-loop gain vs frequency curves intersect at a rate of closure greater than 6 dB/octave, the circuit may be unstable. In the case of the circuit in Fig. 8-14a, the input capacitor C causes a low-frequency roll-off that intersects the open-loop curve at a rate of closure greater than 6 dB/octave as shown in Fig. 8-14b. This means that the circuit will probably be unstable. The problem can be solved if we add two components as shown in Fig. 8-15a. Components R_1 and C_1 cause a 6 dB/octave roll-off with decreasing frequencies, while R_2 and C_2 cause a 6 dB/octave roll-off at higher frequencies, resulting in a closed-loop gain vs frequency curve as shown in Fig. 8-15b. The components R_1, R_F, and C can be selected so that the closed-loop gain vs frequency curve does *not* intersect with the open-loop curve, thus assuring us of stable operation.

Some input voltages V_s and the resulting differentiated outputs V_o are shown in Fig. 8-16.

8-6   Op Amps in Analog Computers

The Op Amp is the fundamental building block of the analog computer. Analog computers are very important and often necessary tools of research engineers and scientists. They can be programmed to solve mathematical models that represent the behavior of some given mechanical or electrical system. In this way, the analog computer, which is basically a collection of Op Amps that can be conveniently wired as needed, can simulate a physical system and show how it will work before the system is actually built. While a thorough or even elementary treatment of analog computers is beyond the scope of this text, we can look at some simple applications of Op Amps as problem-solving and system-simulating components.

If an object is dropped from some height, it will behave in a predictable way that can be expressed in well-established mathematical terms. Generally, if we neglect the effect of air friction and call the distance an object falls s, then its instantaneous velocity v is the derivative of (rate of change in) distance s, that is,

$$v = \frac{ds}{dt}$$

and therefore

$$s = \int v \, dt$$

The instantaneous acceleration a is the derivative of the velocity v and the second derivative of distance s. Thus

$$a = \frac{dv}{dt} = \frac{d^2 s}{dt^2}$$

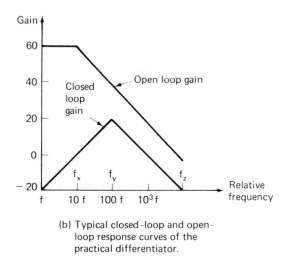

(a) Practical differentiator

$$f_x = \frac{1}{2\pi R_F C_1}$$

$$f_y = \frac{1}{2\pi R_1 C_F}$$

f_z is the frequency at which the open loop-gain is unity

(b) Typical closed-loop and open-loop response curves of the practical differentiator.

Fig. 8-15

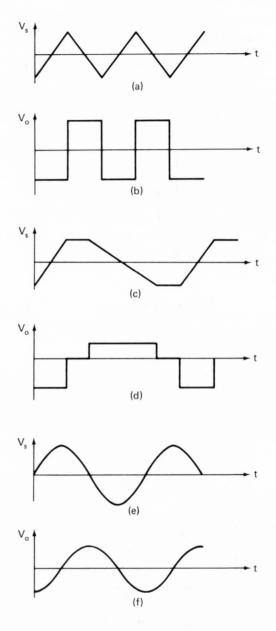

Fig. 8-16 Differentiator's input and resulting output waveforms: input waveform (a) causes output waveform (b); input waveform (c) causes output waveform (d); and input waveform (e) causes output waveform (f).

and therefore

$$v = \int a \, dt$$

If an object starts to fall from rest and is acted on only by gravity, whose acceleration is constant about -32 ft/s^2, then the velocity is

$$v = \int -32 \, dt = -32t$$

and the distance is

$$s = \int -32t \, dt = -16t^2$$

The negative signs on velocity and distance appear because the object is falling, not rising.

This falling object can be simulated by the circuit in Fig. 8-17. Its acceleration a, velocity v, and distance s can be viewed on an oscilloscope or chart recorder at points x, y, and z, respectively. In this case, the constant voltage of -3.2 V is used to represent the constant acceleration of gravity, -32 ft/s^2. The switches S_1 and S_2 are opened simultaneously at t_1 which represents the instant the object is dropped.

Harmonic motion can be simulated with an Op Amp system wired in a closed loop. For example, the spring-mass system shown in Fig. 8-18 will resonate if a vertical force momentarily acts on the mass m. This means that, if the mass m is pulled down against the resisting force of the spring (ky), energy is added to the system and the mass will oscillate after it is released. In the practical case, the oscillations eventually damp out because resistance D of the system consumes all of the energy initially placed into it by the force $F(t)$. If we call the vertical up-and-down motion of the mass y, then this motion can be described with the equation

$$m \frac{d^2v}{dt^2} = F(t) - D \frac{dy}{dt} - ky$$

where: m is the mass,

$F(t)$ is the force, as a function of time, momentarily acting on the mass,

D is the resistance or damping factor of the system, and

k is the force constant of the spring.

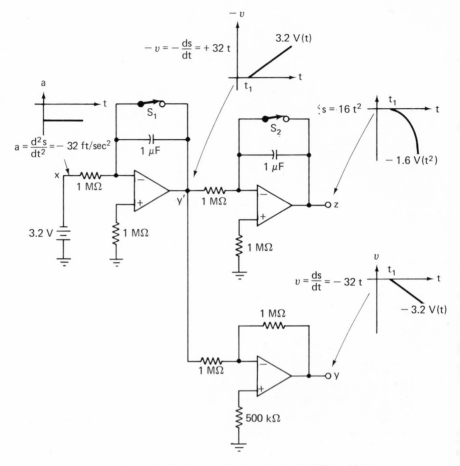

Fig. 8-17 An Op Amp system that simulates a falling object.

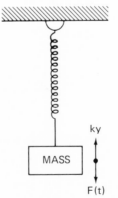

F(t)   **Fig. 8-18** Spring-mass system.

More mass m or a   creases the frequency of the
oscillations. More res   ses the oscillations to damp
sooner. The Op Amp   es this spring-mass system.
The vertical motion y)ut of the second integrator
with an oscilloscope (waveforms that might be
viewed are shown in   sents the instant that the
program starts, that is   ) switches are opened and
represents the instant t

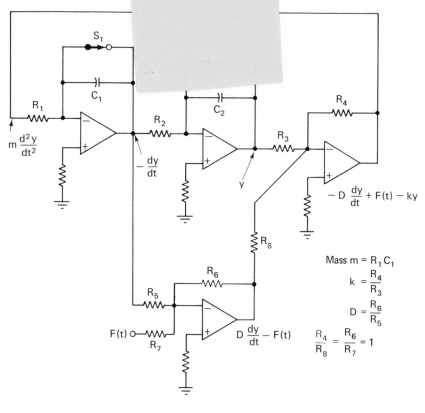

Fig. 8-19   Op Amps programmed to simulate a spring-mass (resonant)
system.

8-7   Op Amps in Voltage Regulators

The Op Amp can be used as a simple low-current dc regulated voltage
source as shown in Fig. 8-21. If the reference voltage V_{ref}, applied to the

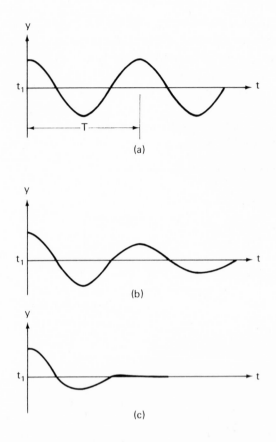

Fig. 8-20 (a) Waveform y when the system resistance (damping factor) D is small. (b) Finite resistance D causes a damped waveform y. (c) Relatively large resistance D causes rapid damping of waveform y. The period T is varied by varying m or k.

noninverting input of the circuit shown in Fig. 8-21a, is constant, the resulting output voltage V_o is relatively constant too. Thus, although the load resistance may vary, the voltage across it is the product of the reference voltage V_{ref} and the closed-loop gain:

$$V_o = A_v V_{ref} \cong \left(\frac{R_F}{R_1} + 1 \right) V_{ref} \tag{8-16}$$

Reliable regulation, however, is limited to load resistances in the range of about $1\ k\Omega$ up to $\infty\ \Omega$ (an open) with general-purpose IC Op Amps. A relatively constant reference voltage can be obtained from a zener (regulator)

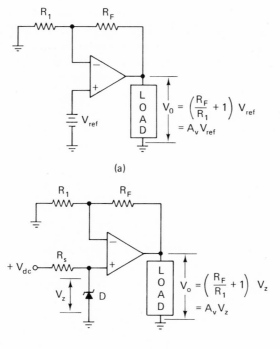

(a)

(b)

Fig. 8-21   Op Amps as low-current-capability voltage regulator.

diode D as shown in Fig. 8-21b. In such circuits, the series resistor R_s is chosen so that its resistance is

$$R_s = \frac{V_{\min} - V_z}{I_z} \qquad (8\text{-}17)$$

where: $V_{\min}$ is the minimum value of the source voltage V_{dc},

V_z is the zener voltage, and

I_z is the zener current which is selected to be in the range of the specified knee current I_{zn} and the test current I_{zt}.

The zener's power rating must exceed

$$P_z = V_z I_z$$

The circuits in Fig. 8-21 can work off a single dc supply voltage where the positive terminal of the supply is connected to the $+V$ pin of the Op Amp. The $-V$ pin of the Op Amp is connected to ground or a common point of the circuit. Such a single supply must be larger than the maximum required output voltage V_o, but not larger than twice the maximum $\pm V$ supply voltage recommended for the Op Amp. By changing the feedback resistance R_F (closed-loop gain), we can change the regulated output V_o, but of course, it is limited to values less than the single supply voltage used to power the Op Amp.

Table 8-1 1-Watt Zener Diodes

JEDEC Type No.	Nominal Zener Voltage V_z @ I_{zt} Volts	Test Current I_{zt} mA	Max Zener Impedance		Max DC Zener Current I_{zm} mA
			Z_{zt} @ I_{zt} Ohms	Z_{zk} @ $I_{zk} = 1.0$ mA Ohms	
1N3821	3.3	76	10	400	276
1N3822	3.6	69	10	400	252
1N3823	3.9	64	9	400	238
1N3824	4.3	58	9	400	213
1N3825	4.7	53	8	500	194
1N3826	5.1	49	7	550	178
1N3827	5.6	45	5	600	162
1N3828	6.2	41	2	700	146
1N3829	6.8	37	1.5	500	133
1N3830	7.5	34	1.5	250	121

EXAMPLE 8-4

If $R_1 = R_F = 10$ kΩ and V_{dc} varies between 20 V and 24 V in a circuit like that in Fig. 8-21b, and if a regulated output V_o of about 10 V is needed, what zener type in Table 8-1 and what resistance R_s should we use? How much power does the zener dissipate?

Answer. The Op Amp is in the noninverting mode as the zener voltage V_z sees it. Therefore, the gain of this circuit is

$$A_v \cong \frac{R_F}{R_1} + 1 = 2$$

Thus in this case, the output voltage is twice the zener voltage. Since the required output is to be about 10 V, the reference or zener voltage must be about 5 V. According to Table 1, a 1N3826 type zener, whose zener voltage is 5.1 V, will work fine. Since this zener's test current I_{zt} is 49 mA, we can use a series resistance

$$R_s = \frac{V_{min} - V_z}{I_z} = \frac{20 \text{ V} - 5.1 \text{ V}}{49 \text{ mA}} \cong 304 \, \Omega \qquad (8\text{-}17)$$

Of course, we can use the nearest available value, 300 Ω or 330 Ω. The actual power dissipated by the zener is

$$P_z = V_z I_z \cong 5.1 \text{ V}(49 \text{ mA}) \cong 250 \text{ mW}$$

Where the load current is to be about 1 A, the circuit in Fig. 8-22a can be used to provide a regulated voltage. In this case a transistor Q, sometimes called a pass element, drops the difference in the unregulated source voltage V_{dc} and the load voltage V_o'. The transistor Q conducts the sum of the load and zener currents. This total current in the emitter of Q is controlled by a much smaller current in its base. Note that the base current is supplied by the Op Amp.*

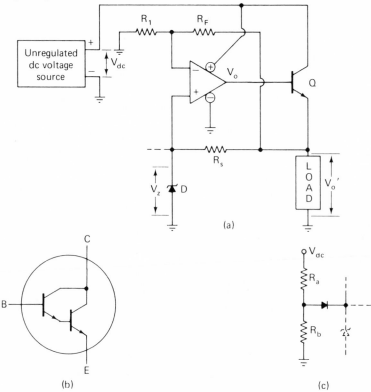

Fig. 8-22 (a) Regulated dc source with intermediate current capability. (b) Darlington pair can be used in place of transistor Q for higher load-current capability. (c) Voltage divider that can be used to assure starting (cause zener to go into avalanche conduction).

---

* A 50-Ω to 200-Ω resistance is sometimes used in series with the base of Q to protect the circuit with excessively low load resistances.

This circuit can be analyzed as follows: A current maintained through the zener provides a well-regulated voltage at the noninverting input. This zener's voltage V_z causes an Op Amp output voltage V_o that is A_v times larger. The voltage V_o is applied to the base of the transistor Q and forward biases its base-emitter junction. For a silicon transistor, this base-emitter drop V_{BE} is typically about 0.7 V. Therefore, the load voltage V'_o, which is the voltage at the emitter, is less than the Op Amp's output V_o by the base-emitter drop V_{BE}, therefore

$$V'_o = V_o - V_{BE} \qquad \text{(8-18a)}$$

or approximately

$$V'_o = V_o - 0.7 \text{ V} \qquad \text{(8-18b)}$$

and often

$$V'_o \cong V_o \qquad \text{(8-18c)}$$

Substituting V_z for V_{ref} in Eq. (8-16) we can show

$$V'_o = A_v V_z \qquad \text{(8-19a)}$$

or

$$V'_o \cong \left(\frac{R_F}{R_1} + 1 \right) V_z \qquad \text{(8-19b)}$$

Apparently then, this circuit's ability to regulate the load voltage V'_o is directly related to the zener's ability to hold a constant drop V_z. Generally, the lower the specified zener dynamic resistance (impedance) Z_{zt}, the better the regulation.

In the circuit of Fig. 8-22a, the drop across R_s is the difference in load and zener voltages, $V'_o - V_z$. Since V_o is regulated, this resistor's drop is quite constant, and therefore, so is the current through it and the zener. A constant current through the zener will cause a constant drop across it regardless of its dynamic impedance Z_{zt}. Thus, the relatively constant zener current, caused by the constant drop across R_s, gives this circuit its good voltage regulating capability.

This circuit's components can be selected as follows: Resistance R_s is selected so that

$$R_s = \frac{V'_o - V_z}{I_z} \qquad \text{(8-20)}$$

where the zener current is selected to be between I_{zn} and I_{zt}; see equation (8-17). The minimum value of the unregulated voltage V_{dc} must be greater than the required load voltage V'_o. To put it another way, the regulated load voltage V'_o must be less than the minimum V_{dc}. The difference in V_{dc} and V'_o appears across the transistor Q. This transistor is selected so that its maximum collector current capability exceeds the maximum current we expect the load to draw, plus the zener current I_z, that is,

$$I_{C(max)} > I_{L(max)} + I_z \qquad (8\text{-}21)$$

Similarly, its power rating must exceed

$$P_{C(max)} = (V_{dc(max)} - V'_o)I_{C(max)} \qquad (8\text{-}22)$$

where $V_{dc(max)}$ is the maximum expected output voltage of the unregulated supply.

The zener voltage V_z and the resistors R_1 and R_F are selected to be consistent with Eq. (8-17a) where, of course,

$$A_v \cong \frac{R_F}{R_1} + 1 \qquad (3\text{-}6)$$

The zener's power rating must exceed its power dissipation as determined with the equation $P_z = V_z I_z$.

As mentioned before, the maximum load current $I_{L(max)}$ of the circuit in Fig. 8-22a is limited to about 1 A using a general-purpose Op Amp. Its load current capability can be significantly increased to several amperes if the transistor Q is replaced with a *Darlington pair*, shown in Fig. 8-22b.

It is possible that when this circuit is initially turned on, the transistor Q may not turn on hard enough to provide a load voltage V'_o that is greater than V_z. In such cases, the zener does not go into avalanche conduction, and the circuit will not regulate the load voltage. A starting voltage can be used across the zener, provided by a voltage divider such as that in Fig. 8-22c. The resistors R_a and R_b are selected so that the drop across R_b exceeds the zener voltage V_z, that is, so that

$$\frac{V_{dc(min)}(R_b)}{R_a + R_b} > V_z \qquad (8\text{-}23)$$

The regulated output voltage V'_o from the circuit in Fig. 8-22 can be made variable, within limits, by replacing R_1 and R_F with a potentiometer. The slide of the potentiometer is connected to the inverting input of the Op Amp.

EXAMPLE 8-5

In the circuit of Fig. 8-22a, suppose that V_{dc} varies between 20 V and 24 V, V_o' is to be regulated at 18.1 V using a 1N3825 type zener, and the load draws from zero to 600 mA. Select the proper ratio of R_F/R_1, the approximate ratio R_a/R_b, and the resistance of R_s. What maximum possible current does the transistor Q draw and what maximum power does it dissipate? How much power does the zener dissipate? Assume that the transistor's $V_{BE} = 0.7$ V.

Answer. Given the required regulated output V_o' of 18.1 V and $V_z = 4.7$ V, the zener voltage of the 1N3825 shown in Table 8-1, we can determine the ratio R_F/R_1—by rearranging Eq. (8-19a). Since

$$A_v \cong \frac{V_o'}{V_z}$$

in this case, we need an

$$A_v \cong \frac{18.1 \text{ V}}{4.7 \text{ V}} = 3.85$$

And since

$$A_v \cong \frac{R_F}{R_1} + 1$$

then

$$\frac{R_F}{R_1} \cong A_v - 1$$

and therefore

$$\frac{R_F}{R_1} \cong 2.85$$

A starting voltage can be provided with a voltage divider such as that in Fig. 8-22c. In this case,

$$\frac{V_{dc(min)}(R_b)}{R_a + R_b} > V_z = 4.7 \text{ V}$$

and therefore

$$\frac{20 \text{ V}(R_b)}{R_a + R_b} > 4.7 \text{ V}$$

from which we can show that, in order to have reliable starting, we must use a ratio $R_a/R_b < 3.26$.

Since Table 8-1 shows that the test current of the 1N3825 is 53 mA, we can use an

$$R_s = \frac{V'_o - V_z}{I_z} = \frac{18.1 \text{ V} - 4.7 \text{ V}}{53 \text{ mA}} \cong 250 \text{ }\Omega$$

The transistor will have to carry as much as

$$I_{C(\text{max})} = I_{L(\text{max})} + I_z \cong 653 \text{ mA}$$

The maximum transistor power dissipation is

$$P_{C(\text{max})} = (24 \text{ V} - 18.1 \text{ V})0.653 \text{ A} \cong 3.86 \text{ W}$$

And the zener's power dissipation is

$$P_z = 4.7 \text{ V}(53 \text{ mA}) \cong 250 \text{ mW}$$

8-8 Active Filters

Filters are used in electrical and electronic applications where certain frequencies must be either attenuated or amplified. In the past, they were constructed primarily with passive components, inductors L and capacitors C. Now with the availability of physically small and low-cost amplifiers, such as IC Op Amps, designers are relying more on active filters.

Twin-T "Notch-Out" Filters. A common RC arrangement frequently used with active filters is the twin-T network shown in Fig. 8-23. Its frequency response is shown in Fig. 8-23b. Note that it greatly attenuates the frequency f_0 which can be selected by proper choice of R and C components.* The frequency f_0 is at the base of the notch in the response curve and its value is

$$f_0 = \frac{1}{2\pi R_1 C_2} \qquad (8\text{-}24)$$

where

$$R_2 = \frac{R_1}{2} \qquad (8\text{-}25)$$

---

* Use metallized polycarbonate or silvered mica capacitors for good temperature stability.

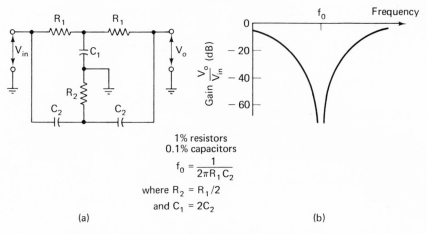

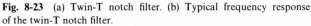

(a)   (b)

Fig. 8-23 (a) Twin-T notch filter. (b) Typical frequency response of the twin-T notch filter.

and

$$C_1 = 2C_2 \qquad (8\text{-}26)$$

This twin-T network has a relatively low Q and a resulting broad bandwidth. Its Q can be significantly multiplied with the use of an Op Amp as shown in Fig. 8-24. The Op Amp's output is fed back to the junction of C_1 and R_2. This raises the effective Q of the twin-T network and results in a much narrower bandwidth, as shown in Fig. 8-24b. The bandwidth can be varied if we vary the amount of feedback to C_1 and R_2 as shown in Fig. 8-25.

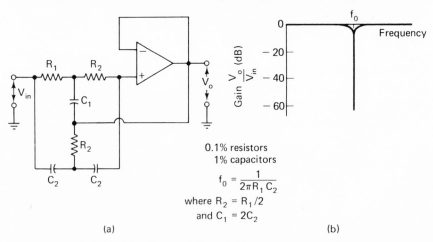

(a)   (b)

Fig. 8-24 (a) Active notch filter. (b) Frequency response of an active notch filter.

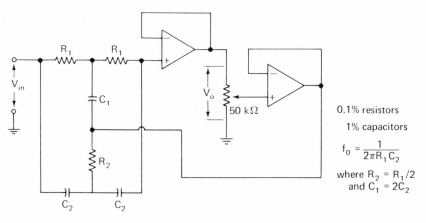

Fig. 8-25   Twin-T notch filter with means of varying its Q.

EXAMPLE 8-6

Referring to the circuit in Fig. 8-25, what frequency is attenuated if $R_1 = 10$ MΩ, $C_1 = 540$ pF, $R_2 = 5$ MΩ, and $C_2 = 270$ pF?

Answer. $f_0 \cong 59$Hz

Twin-T "Notch-In" Filters. The twin-T circuits in Figs. 8-23 through 8-25 "notch-out" (remove) some unwanted frequency f_0 or narrow range of frequencies. The twin-T network is also used to "notch-in" or amplify a wanted frequency or range of frequencies. The circuit in Fig. 8-26 does this. Note in this case that the twin-T network is used in the feedback path and that the feedback is to the inverting input. Since this network acts as a filter to frequency f_0, it causes minimum feedback and consequently maximum circuit gain at the notch frequency f_0. The resulting gain vs frequency response curve is shown in Fig. 8-26b. The notch frequency is still determined by the R and C values of the twin-T, that is, Eqs. (8-24) through (8-26) apply.

The Low-Pass and High-Pass Active Filters. Low-pass filters, as their name implies, have a somewhat flat response at low frequencies, but the response rolls off rather rapidly at higher frequencies. The point where significant roll-off begins can be selected by the designer. A low-pass active filter and its typical response curve are shown in Fig. 8-27. The filter's cutoff frequency f_c is defined as the frequency at which the output signal drops by 3 dB. As shown in Fig. 8-27, the curve rolls off to 3 dB when the ratio of the applied frequency f and the filter's cutoff frequency f_c is 1. Note that if the applied frequency is then doubled, $f/f_c = 2$, the output drops rapidly to about 12 dB.

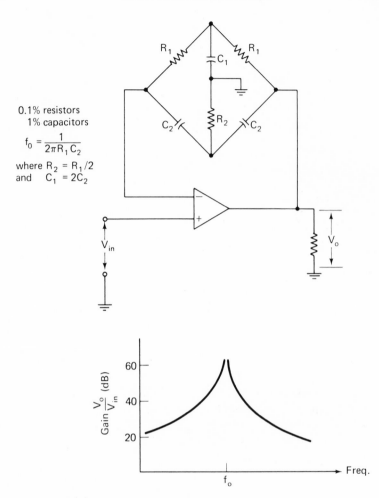

Fig. 8-26   Op Amp and twin-T used to amplify a selected frequency f_0.

In other words, beyond $f/f_c = 1$, the roll-off is roughly 9 dB/octave. After we select f_c, the filter components are related by the following equations:

$$C_1 = \frac{R_1 + R_2}{2\sqrt{2\pi f_c R_1 R_2}} \tag{8-27}$$

$$C_2 = \frac{1}{\sqrt{2\pi f_c (R_1 + R_2)}} \tag{8-28}$$

If we substitute resistors for capacitors and vice versa, the low-pass filter becomes a high-pass filter as shown in Fig. 8-28.

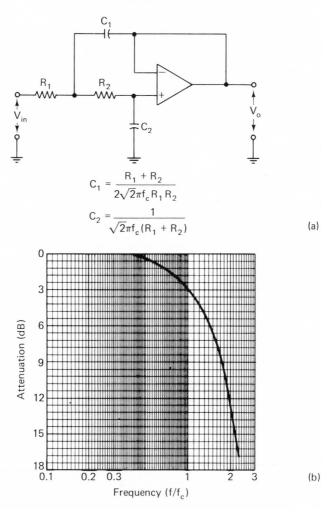

$$C_1 = \frac{R_1 + R_2}{2\sqrt{2}\pi f_c R_1 R_2}$$

$$C_2 = \frac{1}{\sqrt{2}\pi f_c (R_1 + R_2)}$$

(a)

(b)

Fig. 8-27 (a) Low-pass filter, and (b) its typical frequency response.

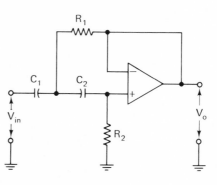

Fig. 8-28 High-pass filter.

REVIEW QUESTIONS

1. Generally, what is the meaning of the term *linear application* of an Op Amp?

2. Since Op Amps are capable of amplifying dc and ac voltages, why should we ever consider using coupling capacitors between Op Amp stages in ac amplifier systems?

3. What is the meaning of and the reason for bootstrapping the input of an amplifier?

4. When using an Op Amp with a single positive power supply, what do we do with the $+V$ and the $-V$ terminals of the Op Amp?

5. If an Op Amp wired as an integrator has a square-wave input voltage applied, what output waveform can we expect?

6. If an Op Amp wired to work as a differentiator has a sawtooth input voltage applied, what output voltage waveform can we expect?

7. What kind of computers use Op Amps as basic building blocks?

8. An Op Amp with a reference voltage can provide regulated voltage to varying loads, provided they draw moderate currents, well under an ampere maximum. What can be added to such a power supply to increase its current capability?

9. What advantages do active filters have compared with passive types?

10. What network of resistors and capacitors can be used with an Op Amp to either reject or select a narrow band of frequencies?

11. In the circuit of Fig. 8-29, what is the purpose of the 240-kΩ resistor?

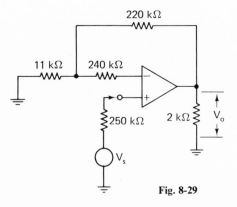

Fig. 8-29

PROBLEMS

1. What is the gain V_o/V_s of the circuit in Fig. 8-29?

2. In the circuit of Fig. 8-29, if the 11-kΩ resistor is replaced with an 82-kΩ resistor, (a) what is the gain V_o/V_s of the circuit, and (b) with what value should the 240-kΩ resistor be replaced if the circuit stability is to remain close to what it was?

3. In the circuit of Fig. 8-30, (a) what is the gain V_o/V_s of the circuit, and (b) what value of resistance R should we use to minimize instability and output offset voltage?

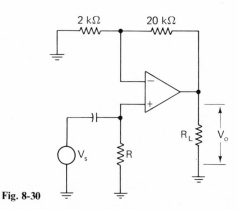

Fig. 8-30

4. If the Op Amp in the circuit of Fig. 8-30 is a type 741 and $R = 1$ MΩ, roughly what maximum output offset voltage, caused by bias currents, could we expect at room temperature?

5. In the circuit of Fig. 8-31, what total gain V_o/V_s can we expect at frequencies at which the reactances of all capacitors are negligible if the potentiometer is adjusted to 0 Ω?

6. At frequencies that cause the reactances of the capacitors to be negligible in the circuit of Fig. 8-31, what gain V_o/V_s can we expect if the potentiometer is adjusted to maximum resistance?

7. Referring to the circuit in Fig. 8-5, if $R_1 = 10$ kΩ, $R_2 = 200$ kΩ, $R_F = 200$ kΩ, $R = 52$ Ω, $+V = 15$ V, and the zeners are both 1N3828 types, (a) what dc voltage can we expect at pin 6, and (b) what is the gain V_o/V_s of the circuit? Assume that the reactances of all capacitors are negligible.

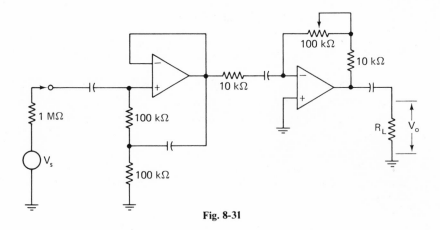

Fig. 8-31

8. In the circuit described in Problem 7, (a) what is the current in resistor R?, and (b) what approximate power is dissipated in each of the zeners if the Op Amp draws 4 mA?

9. In the circuit described in the Problem 7, if the input V_s is sinusoidal with a peak of 200 mV, what are the instantaneous positive and negative voltage peaks (a) at pin 6 and (b) at the right of the capacitor C_2?

10. In the circuit of Fig. 8-32, if $R_1 = R_2 = R_3 = 1\ k\Omega$ and if the Op Amp was initially nulled, (a) what is its output voltage V_o, and (b) what value of R_4 should we use to minimize drift?

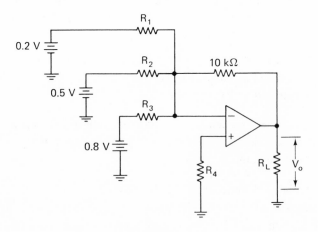

Fig. 8-32

11. If $R_1 = 1\,\text{k}\Omega$, $R_2 = 2\,\text{k}\Omega$, and $R_3 = 4\,\text{k}\Omega$ in the circuit of Fig. 8-32, (a) what is its output voltage V_o, and (b) what value of R_4 will minimize drift? Assume that the Op Amp was initially nulled.

12. The three input resistors R are to be equal in the circuit of Fig. 8-33. (a) In order to minimize drift, what value should each of these input resistors be? (b) What is the voltage at the noninverting input, and (c) what is the voltage V_o? Assume that the Op Amp was nulled.

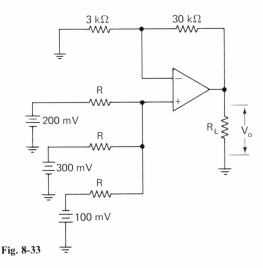

Fig. 8-33

13. In the circuit of Fig. 8-10, if $R = 100\,\text{k}\Omega$, $C = 100\,\mu\text{F}$, and the applied voltage V_s has the waveform shown in Fig. 8-11, what is the output voltage at (a) $t = 1.1$ second and at (b) $t = 5.1$ seconds. Assume that initially the Op Amp was nulled and that the capacitor's voltage was zero.

14. In the circuit described in Problem 13, how long will it take for the amplifier to saturate after the 8-V step voltage is applied, if the output saturation voltages are $+16\,\text{V}$ and $-16\,\text{V}$?

15. What are the voltages at points y', y, and z in the circuit of Fig. 8-17 at the instant 0.5 seconds after the switches are open?

16. What are the voltages at points y', y, and z in the circuit of Fig. 8-17 at the instant 2 seconds after the switches are open?

17. What is the load voltage V_o in the circuit of Fig. 8-34 if the zener is a 1N3821 and $R_1 = R_F = 10\,\text{k}\Omega$? If the zener current is to be 70 mA and $V_{\text{dc}} = 15$ V, what value of resistance R should we use?

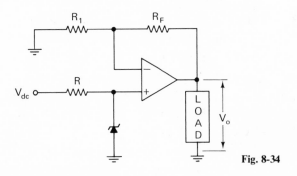

Fig. 8-34

18. What is the load voltage V_o in the circuit of Fig. 8-34 if the zener is a 1N3822 and if $R_1 = 10\,k\Omega$, $R_F = 30\,k\Omega$? If $V_{dc} = 20\,V$ and the zener is to draw about 70 mA, what value of resistance R should we use?

19. If in the circuit of Fig. 8-22, V_{dc} varies between 20 V and 28 V, $R_1 = 10$ $k\Omega$, $R_F = 20\,k\Omega$, and the zener is a 1N3823 type, what is the voltage V_o' and the drop across R_s? Assume that the transistor's $V_{BE} = 0.7\,V$.

20. If in the circuit of Fig. 8-22, V_{dc} varies between 24 V and 30 V, $R_1 = 10$ $k\Omega$, $R_F = 40\,k\Omega$, and the zener is a 1N3822 type, what is the voltage V_o' and the drop across R_s? Assume that the transistor's $V_{BE} = 0.7\,V$.

21. In the circuit described in Problem 19, what are the maximum power values dissipated (a) by the transistor and (b) by R_s if the load draws up to 0.8 A and the zener draws 65 mA?

22. In the circuit described in Problem 20, what are the maximum power values dissipated (a) by the transistor and (b) by R_s if the load draws up to 1 A and the zener draws 70 mA?

23. In the circuit of Fig. 8-25, if $R_1 = 20\,k\Omega$, $R_2 = 10\,k\Omega$, and the circuit is to "notch-out" approximately 400 hz, what values of C_1 and C_2 should we use?

24. If in the circuit of Fig. 8-26, $R_1 = 40\,k\Omega$, $R_2 = 20\,k\Omega$, and the circuit is required to "notch-in" approximately 20 hz, what values of C_1 and C_2 should we use?

25. What is the approximate cutoff frequency f_c in the circuit of Fig. 8-27 if $C_1 = 940\,pF$, $C_2 = 470\,pF$, and $R_1 = R_2 = 24\,k\Omega$?

CHAPTER 9

NONLINEAR
APPLICATIONS
OF OP AMPS

In some applications, Op Amps are not required to work as linear amplifiers. Instead, they are required to abruptly switch from one output voltage level to another even though the input voltage changes are gradual. Occasionally the output voltage swing of an Op Amp is to be kept within specific limits. In such cases, Op Amps are used with externally wired components that clip output signals which attempt to swing beyond the predetermined limits. Beyond these, there are limitless possible nonlinear applications of Op Amps. The analyses of a few in this chapter will give us insight into other possibilities.

9-1   Voltage Limiters

The output swing of a general-purpose Op Amp is often too large for the inputs of some circuits. Digital circuits, for example, usually require specific levels of inputs that are not, without modification, available from the output of a typical Op Amp. Several output-voltage-limiting circuits are shown in Figs. 9-1 through 9-5.

In the circuit of Fig. 9-1, the zener diodes D_1 and D_2 limit the peak-to-peak value of the output voltage V_o. The zener voltage V_{z_1} of D_1 is approximately equal to the maximum possible positive value of V_o, i.e., we cannot drive the output V_o more positive than about the V_{z_1} voltage. The zener voltage V_{z_2} of D_2 similarly is about equal to the maximum negative value of V_o.

More specifically, as shown by the transfer characteristic in Fig. 9-1b, the output V_o levels off (is clipped) at a voltage $V_{z_1} + V_F$ as it swings positively. On the other hand, as the output V_o swings negatively, it is clipped at $-(V_{z_2} + V_F)$. The voltage V_F is the voltage drop across the *forward*-biased

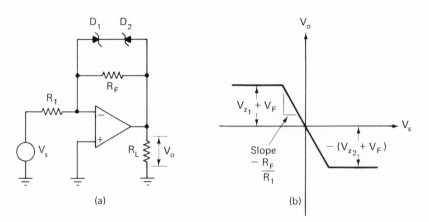

Fig. 9-1 (a) Positive and negative voltage clipper and (b) its transfer characteristics.

zener and is typically about 0.7 V. Of course, when V_o swings positively, D_1 is reverse biased and D_2 is forward biased; when V_o swings negatively, D_1 and D_2 are forward and reverse biased, respectively.

EXAMPLE 9-1

In the circuit of Fig. 9-1, suppose that each zener has a 6.3-V zener voltage and a 0.7-V forward drop, and that $R_1 = 1 \text{ k}\Omega$ and $R_F = 20 \text{ k}\Omega$. Describe the output waveforms with each of the following sine-wave inputs:

(a) $V_s = 0.3$ V peak,
(b) $V_s = 0.6$ V peak, and
(c) $V_s = 3$ V peak.

Answer. The gain A_v of this stage is about $-R_F/R_1 = -20$, and its output peaks are limited to

$$V_{z_1} + V_F = 6.3 + 0.7 = 7 \text{ V}$$

on positive alternations and to

$$-(V_{z_2} + V_F) = -(6.3 + 0.7) = -7 \text{ V}$$

on the negative alternations.

(a) When $V_s = 0.3$ V peak, $V_o = A_v V_s = -20(0.3) = -6$ V peak. This output V_o does not force either zener into zener (reverse) conduction, and therefore the output waveform is unclipped and sinusoidal.

(b) When $V_s = 0.6$ V peak, the output would peak to $-20(0.6) =$ -12 V if the zeners were not across R_F. Since they are, V_o gets clipped at 7 V on positive and negative alternations.

(c) When $V_s = 3$ V peak, the output would peak to $-20(3) = -60$ V if limiting factors were not present. Even without the zeners, the general-purpose Op Amp will saturate if we attempt to drive it this hard. With the zeners, the output is clipped at about 7 V on each alternation, and the output V_o has the appearance of a square wave.

The zener and rectifier diodes in the circuit of Fig. 9-2 limit the swing of the output voltage V_o in a positive direction only, as shown by its transfer characteristic in Fig. 9-2b. This means that, if V_o attempts to rise above the

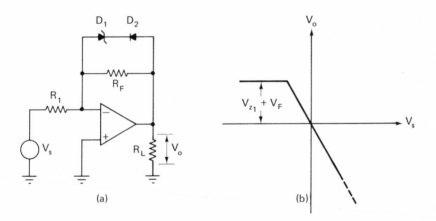

Fig. 9-2　(a) Positive voltage clipper and (b) its transfer characteristics.

voltage $V_{z_1} + V_F$, the zener D_1 goes into zener (avalanche) conduction and the waveform of V_o is clipped. The negative alternations are not clipped unless the Op Amp is driven into negative saturation. The circuit in Fig. 9-3 is similar except that its zener and rectifier diodes limit the output V_o to $-(V_{z_2} + V_F)$ when it swings in the negative direction. V_o can swing positively to the point where the Op Amp positively saturates.

If only a single zener is used with no rectifier, as in Figs. 9-4 and 9-5, and a sine-wave input V_s is applied, the swing in the output V_o is limited by the zener voltage on half of the alternations and by the zener's forward drop V_F on the remaining alternations.

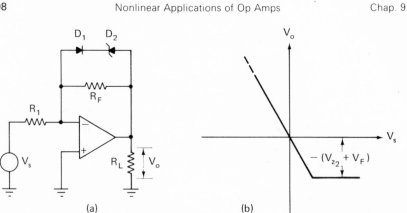

Fig. 9-3 (a) Negative voltage clipper and (b) its transfer characteristics.

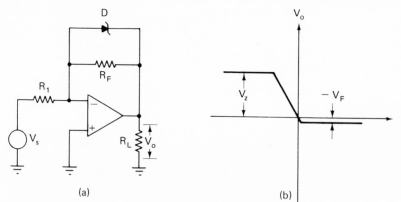

Fig. 9-4 (a) Half-wave rectifier with limited positive output and (b) its transfer characteristics.

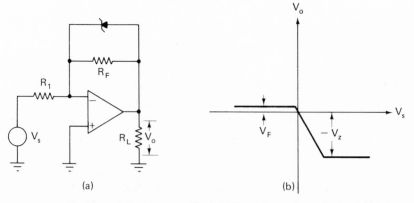

Fig. 9-5 (a) Half-wave rectifier with limited negative output and (b) its transfer characteristics.

EXAMPLE 9-2

Referring to the circuits in Figs. 9-1 through 9-5, suppose that in each, $R_1 = 1\,k\Omega$, $R_F = 10\,k\Omega$, $V_z = 4\,V$ of each zener, and that the drop across each forward-biased diode and zener is negligible. Sketch the output V_o of each circuit if a 2-V peak, 60-Hz sine-wave input voltage V_s is applied to its input.

Answer. See Fig. 9-6a. This waveform is the input V_o applied to each circuit. Each circuit's gain $A_v = -R_F/R_1 = -10$. Thus the outputs V_o are out of phase with V_s and they *tend* to peak at 20 V on positive and negative alternations. Clipping occurs however. The waveform shown in Fig. 9-6b is the output V_o of the circuit in Fig. 9-1. The waveform in c of the figure is the output V_o of the circuit in Fig. 9-2. The waveform in Fig. 9-6d is the output V_o of the circuit in Fig. 9-3. In Fig. 9-6e, the waveform is the output V_o of the circuit in Fig. 9-4. And the waveform in Fig. 9-6f is the output V_o of the circuit in Fig. 9-5.

9-2   The Zero-crossing Detector

When used in open loop, the Op Amp is very sensitive to changes in input voltages. In fact, fractions of millivolts can easily drive the Op Amp into saturation when no feedback is used. This feature is an advantage in some applications. For example, in Fig. 9-7a, the circuit works as a zero-crossing detector or a sine-wave to square-wave converter. As shown in its output V_o vs input V_s waveforms in Fig. 9-7b, the output V_o swings and saturates negatively when the input V_s passes through zero in the positive direction. On the other hand, when V_s passes through zero negatively, the Op Amp's output V_o is driven into positive saturation.

If the peak of the input voltage V_s exceeds a volt or so, the input diodes D_1 and D_2 protect the Op Amp. Of course, if the Op Amp is a type that is input-protected, which means that the equivalent of diodes D_1 and D_2 are built internally, the input diodes are usually unnecessary. If the output swing between $V_{o(max)}$ and $-V_{o(max)}$ is excessive, as it would be for many digital circuit inputs, either or both of the zeners D_3 or D_4 can be used to clip V_o at whatever limits are necessary. The resistance R_1 limits the current through the input protection diodes. If $R_1 = 1\,M\Omega$, V_s values up to 1 kV will force only about 1 mA through these diodes. A large R_1, however, adds to offset errors. Therefore, if the output of this circuit is to switch as precisely as

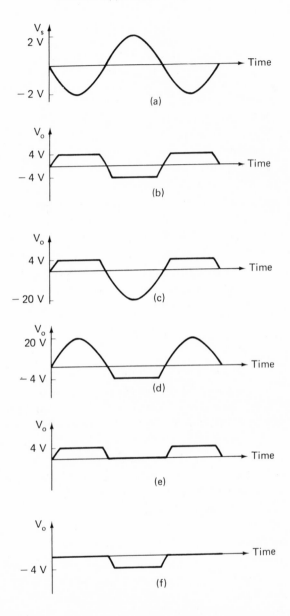

Fig. 9-6 If (a) is the input waveform V_o, $R_F/R_1 = 10$ and $V_z = 4$ V, then (b) is the output V_o of the circuit in Fig. 9-1, (c) is the output V_o of the circuit in Fig. 9-2, (d) is the output V_o of the circuit in Fig. 9-3, (e) is the output V_0 of the circuit in Fig. 9-4, and (f) is the output V_o of the circuit in Fig. 9-5.

possible at the instant voltage V_s passes through zero, a resistance equal to R_1 should be placed between the noninverting input and ground. Such a resistor reduces offset problems and is more necessary when large values of R_1 are used and if the input bias current is relatively large as it is with general-purpose Op Amps.

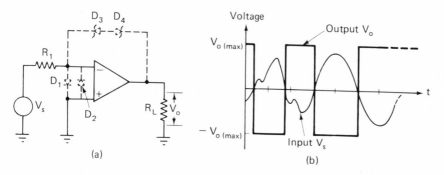

Fig. 9-7 (a) The zero-crossing detector and (b) its typical waveforms. D_1 and D_2 are input-protecting diodes and D_3 and D_4 are output-limiting zeners.

9-3 Op Amps As Comparators

The comparator, as its name implies, compares two voltages. One is usually a fixed reference voltage, V_R, and the other a time-varying signal voltage which is often called an *analog* voltage V_A. The Op Amp comparator circuit is very similar to the zero-crossing detector discussed in the previous section, except that a reference voltage V_R is used between one input and ground on the comparator. See Fig. 9-8. This causes the Op Amp's output to swing from $V_{o(max)}$ to $-V_{o(max)}$, or vice versa, as the analog voltage V_A passes through the reference voltage value V_R. This reference V_R can be either positive or negative with respect to ground, and of course, its value and polarity determine the V_A voltage that causes the output to switch. Note in Fig. 9-8b that, with the analog input waveform, the output has the waveform in Fig. 9-8c or d, depending on whether V_R is positive or negative, respectively. Obviously, the amplitude of V_A must be large enough to pass through V_R if the switching action is to take place. The circuit in Fig. 9-8 is an inverting type. A noninverting Op Amp comparator and its typical waveforms are shown in Fig. 9-9. A resistor $R_2 = 200\,\Omega$ or so and one or both zeners D_3 or D_4 are used if necessary to keep the output swing within required limits.

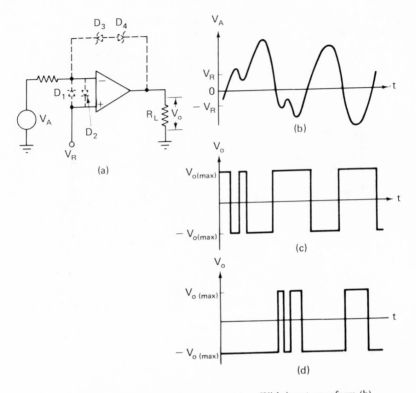

Fig. 9-8 (a) An Op Amp as a comparator. With input waveform (b), the output can be waveform (c) with a positive reference voltage, or waveform (d) if the reference is negative.

If the switching action of the comparator must be very fast, the general-purpose Op Amp's limited slew rate may make it unsuitable. High slew rate Op Amps and IC packages, made specifically to work as comparators, are better suited for high-speed switching.

9-4   An Absolute Value Output Circuit

The output signal V_o of the circuit in Fig. 9-10 can swing positively only, regardless of the polarity of the input signal V_s. Though the input V_s swings through positive and negative values, the output V_o will change proportionally but will vary between zero and positive values. An input V_s and a resulting output V_o for this circuit are shown in Fig. 9-10b and c.

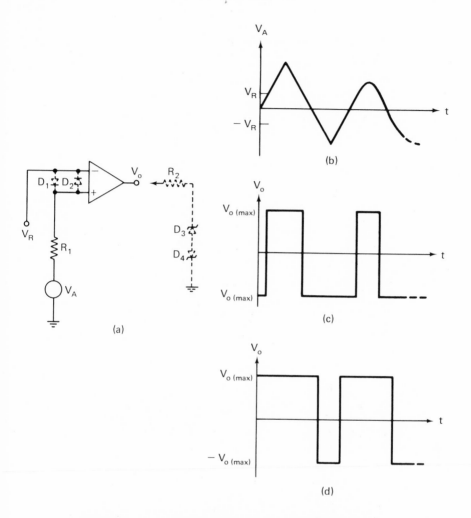

Fig. 9-9 (a) Noninverting Op Amp comparator. With the input wave-form (b), the output can be waveform (c) if the reference voltage is positive, or waveform (d) if the reference is negative.

When V_s is positive, diode D_2 is forward biased, and due to the divider action of the equal resistors R_a and R_b, the signal at input 2 is $V_s/2$. Since D_1 is reverse biased, the equal resistors R_F and R_1 form a simple feedback network of a noninverting amplifier, and therefore the output is

$$V_o = A_v \frac{V_s}{2} \cong \left(\frac{R_F}{R_1} + 1 \right) \frac{V_s}{2} \cong V_s$$

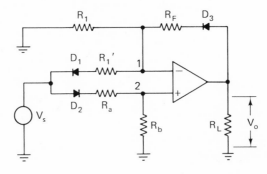

(a) $R_1 = R_1' = R_F = R_a = R_b$

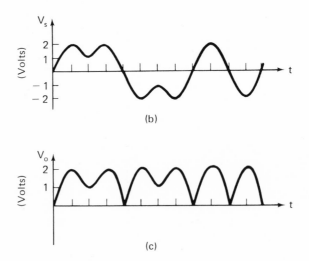

(b)

(c)

Fig. 9-10 (a) Absolute value output circuit. With input waveform (b) its output has waveform (c).

On the other hand, when V_s is negative, D_1 is forward biased and D_2 is reverse biased. Since the left side of R_1 is grounded and now the right side is virtually grounded, it draws negligible current and drops out of consideration as far as circuit gain is concerned. Now the output is

$$V_o = A_v V_s \cong -(R_F/R_1')V_s \cong -V_s$$

Therefore, if resistors R_1, R_1', R_F, R_a, and R_b are all equal, this circuit's gain is either unity or minus unity, depending on the polarity of the input V_s.

Although the voltage gain is small, this circuit is capable of considerable power gain, that is, V_s can have large internal resistance while the load resistance R_L is relatively small.

9-5   The Op Amp As a Small-signal Diode

By themselves, solid-state diodes are not useful for rectification of small signals. This is due to the fact that silicon diodes require about 0.7 V forward bias V_F, and germanium diodes require about 0.3 V forward bias, before any significant conduction occurs. Signals with small amplitudes, therefore, cannot directly drive solid-state diodes into conduction. Of course, then, such signals cannot be rectified simply.

Signals with peak values down to a few millivolts can be rectified with the circuit in Fig. 9-11. It is the high open-loop gain A_{VOL} of the Op Amp

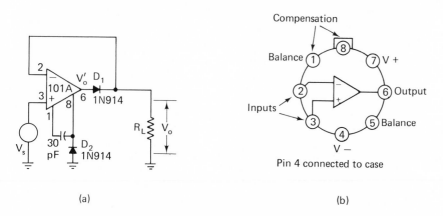

(a)   (b)

Fig. 9-11   (a) An Op Amp connected to work as a small-signal rectifier; (b) metal-case pin designation of the type 101 Op Amp.

that makes this possible. For example, if a small positive signal V_s is applied (a few millivolts peak), the output V'_o at pin 6 *tends* to rise to a value that is A_{VOL} times larger. As V'_o rises, however, it easily reaches the forward bias drop V_F of the diode D_1, causing it to conduct. Thus the output V_o across the load R_L is less than V'_o by the diode's drop, V_F. This final output V_o is fed back to the inverting input and therefore must essentially be equal to the input signal V_s. That is, because there is virtually no potential difference between the inverting and noninverting inputs of high-gain Op Amps, the output V_o

must follow the input V_s for positive values of V_s down to fractions of milli-volts. With negative values of V_s, the diode D_1 is reverse biased and prevents current flow into the load R_L. V_o therefore remains zero.

9-6   The Op Amp as a Sample-and-hold Circuit

The function of the sample-and-hold circuit is somewhat explained by its name. It usually is used to read (sample) an input signal V_s and hold its instantaneous value for a period of time t_H. An Op Amp sample-and-hold circuit is shown in Fig. 9-12. The MOSFET* serves as a switch that is effec-tively opened or closed by the presence or absence of a control voltage V_c on its

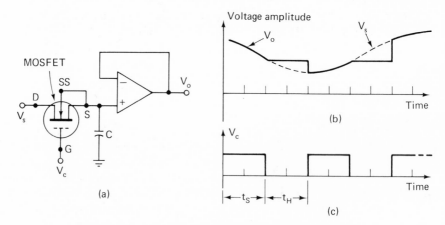

Fig. 9-12   (a) A sample-and-hold circuit with output voltage V_o vs input V_s (b) resulting from the applied control voltage V_c waveform (c).

gate G, that is, a positive pulse V_c applied to the gate G of the enhancement-mode MOSFET causes it to become conductive between its drain D and source S leads. This allows the signal V_s to charge the capacitor C. In fact, the voltage across the capacitor essentially follows V_s when V_c is applied. The time periods when pulses V_c are applied are called *sample* periods t_S. The times when gating pulses V_c are not applied are called *hold* periods t_H. During hold periods t_H, the MOSFET is nonconductive, and the charge in and the voltage across the capacitor C are held constant. The Op Amp's output is usually read or observed during such times.

---

* The MOSFET is a Metal-Oxide Semiconductor Field Effect Transistor.

The sample-and-hold circuit can be used to provide a steady voltage into a device that cannot process a varying signal. An analog-to-digital (A/D) converter might be such a device. Waveforms V_o and V_s in Fig. 9-12b are output and input waveforms of the sample-and-hold circuit when the control voltage V_c has the waveform shown in Fig. 9-12c.

If the times t_S and t_H are short compared with the time it takes the signal V_s to significantly change, successive readings of V_o at instances in t_H give a close approximation of the input waveform. Of course, since the capacitor's function is to hold a constant voltage for periods of time, it should be a low-leakage type, preferably one with a polycarbonate, polyethylene, or teflon dielectric.

PROBLEMS

1. Sketch the waveform of V_o in the circuit of Fig. 9-1 if V_z of each zener is 6 V, $R_F = 100 \text{ k}\Omega$, $R_1 = 10 \text{ k}\Omega$, and the input signal V_s is sinusoidal with a peak of 1 V. Assume that the Op Amp has negligible offset and that the forward drops of both diodes are negligible.

2. In the circuit described in Problem 1, sketch the waveform of V_o if D_2 is replaced with a zener whose $V_z = 3$ V.

3. Sketch the waveform of V_o in the circuit of Fig. 9-2 if V_z of the zener is 4 V, $R_F = 220 \text{ k}\Omega$, $R_1 = 5 \text{ k}\Omega$, and the input signal V_s is sinusoidal with a peak of 250 mV. Assume that the Op Amp is nulled and that the forward drops across both diodes are negligible.

4. Sketch the waveform of V_o in the circuit described in Problem 3 if the 5-kΩ resistor is replaced with 10 kΩ.

5. In the circuit described in Problem 3, sketch its output V_o waveform if its diode D_2 is replaced with a short.

6. Referring again to the circuit described in Problem 3, sketch the waveform of V_o if either D_1 or D_2 becomes an open.

7. If the zener diode D in the circuit of Fig. 9-4 has a $V_z = 10$ V and the input V_s is sinusoidal with a 5-mV peak, what maximum ratio of R_F/R_1 can we use and still prevent clipping of the output signal's positive alternations?

8. Sketch the waveform of V_o that we could expect at the output of the circuit in Fig. 9-5 if its zener's $V_z = 8$ V, $R_F = 220 \text{ k}\Omega$, $R_1 = 1.1 \text{ k}\Omega$, and if V_s is sinusoidal with a peak of 80 mV.

9. The Op Amps in Fig. 9-13 have output limits of $+15$ V and -15 V.
With the input voltage V_A shown, what are the output voltages V_{o_1}, V_{o_2}, and
V_{o_3} at instants (a) t_2, (b) t_3, and (c) t_5? A_{VOL} of each Op Amp is extremely
large.

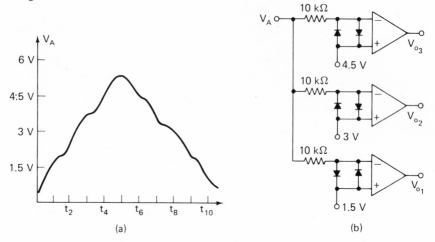

(a)   (b)

Fig. 9-13   Comparator circuit (b) with analog input voltage V_A (a)
applied.

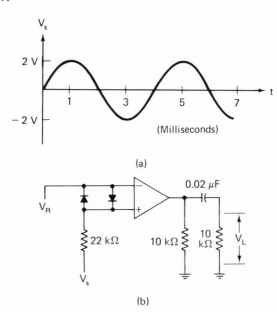

(a)

(b)

Fig. 9-14

10. In the circuit and input described in Problem 9, what are the outputs V_{o_1}, V_{o_2}, and V_{o_3} at times (a) t_7, (b) t_9, and (c) t_{10}? Assume that the open-loop gain A_{VOL} of each Op Amp is extremely large.

11. Sketch the load voltage V_L waveform of the circuit in Fig. 9-14b where the waveform in Fig. 9-14a is the input signal V_s, and $V_R = 0$ V to ground.

12. Sketch the load voltage V_L waveform of the circuit in Fig. 9-14b, where Fig. 9-14a shows the input signal V_s, and $V_R = +1.4$ V dc to ground.

CHAPTER 10

OP AMP
SIGNAL GENERATORS

Op Amps can be wired to serve as signal generators capable of a variety of output waveforms. Square waves, triangular waves, sawtooth waves, and sine waves are readily available, to name the more useful waveforms. In this chapter, we will see a few basic Op Amp signal generators and methods of selecting their externally wired components. The values of these components, and the ways in which they are connected, determine the output waveforms and their frequencies.

10-1   The Square-wave Generator

A simple Op Amp square-wave generator is shown in Fig. 10-1. Its output repetitively swings between positive saturation $+V_{o(\max)}$ and negative saturation $-V_{o(\max)}$, resulting in the square-wave output shown. The time period T of each cycle is determined by the time constant of the components R and C and by the ratio R_a/R_b. This circuit's operation can be analyzed as follows: At the instant the dc supply voltages, $+V$ and $-V$, are applied, zero volts of the initially uncharged capacitor C is applied to the inverting input 1, that is, input 1 is initially grounded. At the same instant, however, a small positive or negative voltage V_b appears across R_b, and this voltage is applied to the noninverting input 2. Voltage V_b initially appears because a positive or negative output offset voltage V_{oo} exists, even if no differential input voltage is applied to inputs 1 and 2. Thus the resistors R_a and R_b form a voltage divider, and a fraction of the Op Amp's output voltage is dropped across R_b. Since the inverting input 1 is initially grounded through the uncharged capacitor C, all of the voltage V_b initially appears across the inputs 1 and 2. Even if V_b is small, it will start to drive the Op Amp into saturation, that is, if the output offset V_{oo} is positive, the voltage V_b at the noninverting input 2

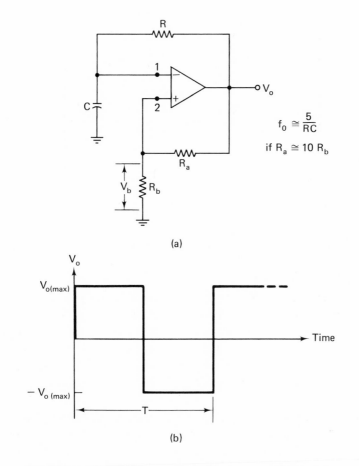

$$f_0 \cong \frac{5}{RC}$$

if $R_a \cong 10 R_b$

(a)

(b)

Fig. 10-1 (a) Square-wave generator. (b) output waveform of the square-wave generator at low frequencies f, where $f = 1/T$.

is positive. This V_b is initially amplified by the Op Amp's open-loop gain A_{VOL} and drives the output to its limit $V_{o(max)}$, that is, to positive saturation. The rise to $V_{o(max)}$ is at the slew rate of the Op Amp. With the Op Amp saturated, the capacitor charges through resistor R. If the resistor R and capacitor C formed a simple RC circuit, the capacitor's voltage V_c would eventually rise to $V_{o(max)}$. In this case, however, voltage V_c can rise only to a value slightly more positive than V_b. That is, as V_c rises and becomes a little more positive than V_b, the inverting input 1 becomes more positive than the noninverting input 2, and this drives the output to its negative limit $-V_{o(max)}$. See Fig. 10-2. After the Op Amp's output saturates at $-V_{o(max)}$, a fraction of

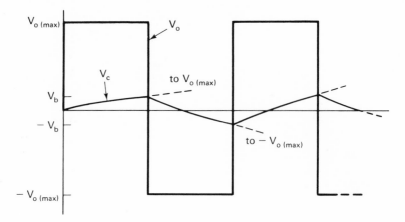

Fig. 10-2 Typical output voltage V_o and capacitor voltage V_c waveforms of the square-wave generator.

this voltage is dropped across R_b. Thus input 2 becomes much more negative than input 1 and holds the Op Amp in negative saturation, at least for a while. The capacitor C now proceeds to discharge and recharge in the negative direction as shown in Fig. 10-2. Now when the capacitor's voltage becomes more negative than $-V_b$, the inverting input 1 becomes more negative than input 2, and the output is driven back to $+V_{o(max)}$ to start another cycle.

The *sum* of the resistances R_a and R_b is not critical. It can be selected to be in a broad range, say 10 kΩ to 1 MΩ. A change in their *ratio*, however, does affect the circuit's output frequency f. Generally, if the ratio $R_a/R_b \cong 10$, then the period of each cycle is

$$T \cong 0.2RC \tag{10-1}$$

and the number of cycles per second is

$$f \cong \frac{5^*}{RC} \tag{10-2}$$

The smaller the RC product, the faster the capacitor C charges to the voltage across R_b, and the higher the output frequency. Therefore the resistor R in Fig. 10-1 can be a frequency-selecting potentiometer.

This square-wave generator's output frequency is limited by the slew rate of the Op Amp. That is, if we attempt to operate it at relatively high fre-

---

* This equation's accuracy deteriorates with larger bias currents and R/C ratios.

quencies, the period of each cycle T is not much larger than the time it takes the output to rise to $V_{o(max)}$ from $-V_{o(max)}$ and vice versa. This causes the generator's output to become trapezoidal or even triangular at higher frequencies.

The peak-to-peak output capability of this square-wave generator can be reduced by means of reduced dc supply voltages or with back-to-back zeners as shown in Fig. 10-3. The zeners will reduce ringing* as well.

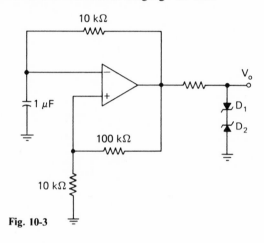

Fig. 10-3

EXAMPLE 10-1

Refer to the circuit of Fig. 10-3. What is its output frequency and the peak-to-peak output voltage V_o if each of the zeners has a 4-V zener voltage V_z? Assume that the drop across each zener is 0 V when it is forward biased.

Answer. $f \cong 5/RC = 5/10^{-2} = 500$ Hz. $V_o = 8$-V peak to peak.

10-2   The Triangular-wave Generator

In Section 8-4 we learned that an integrator's output waveform is triangular if its input is a square wave. See Fig. 8-13a. Apparently, then, an integrator following a square-wave generator, such as in Fig. 10-4, serves as a triangular-wave generator. Since changes in resistance R change the

---

* Ringing refers to an output voltage that oscillates about its eventual steady state after a sudden change.

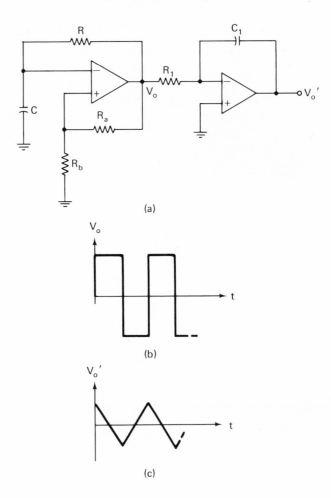

Fig. 10-4   (a) Triangular-wave generator, (b) output waveform of the first Op Amp, and (c) output of the second Op Amp.

frequency of the square-wave generator's output, the output of the integrator is similarly affected. Therefore, if the resistance R is increased or decreased, the frequency of the triangular wave will decrease or increase, respectively. The amplitude of the triangular wave can be controlled somewhat by resistance R_1. Larger or smaller values of R_1 will reduce or increase the output amplitude of the integrator. As with the simple square-wave generator, the output frequency is limited by the Op Amps' slew rates.

10-3   The Sawtooth Generator

A sawtooth waveform differs from the triangular waveform in its unequal rise and fall times. The sawtooth may rise positively many times faster than it falls negatively, or vice versa. The circuit in Fig. 10-5 is a sawtooth generator. The first stage is called a *threshold detector*. Its output V_o will swing from its $V_{o(max)}$ to $-V_{o(max)}$ when the decreasing output V_o' of the integrator becomes negative enough to pull point x slightly negative with respect to ground. Note the waveforms in Fig. 10-5b. On the other hand, the threshold

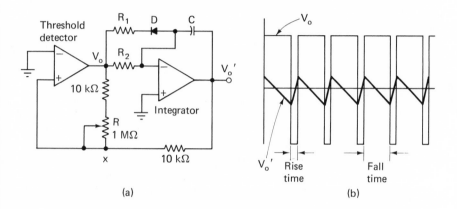

Fig. 10-5   (a) The sawtooth generator and (b) its waveforms if $R_2 > R_1$.

detector's output swings from $-V_{o(max)}$ to $+V_{o(max)}$ when the rising voltage V_o' lifts point x slightly positive with respect to ground. The decrease of V_o' in the negative direction takes longer than its rise in the positive direction because the rate at which the capacitor C charges changes as the polarity of V_o changes. That is, when V_o is saturated negatively, the capacitor C charges mainly through R_1 and the forward-biased diode D. The time constant R_1C is made relatively short if R_1 is made significantly smaller than R_2. When V_o is saturated positively, the capacitor C charges through R_2 more slowly because the time constant R_2C is relatively long. The values of R_1 and R_2 largely dictate the frequency of the output V_o', while their ratio R_1/R_2 determines the ratio of the rise and fall times. If we reverse the diode D, the rise time of V_o' becomes larger than the fall time.

10-4   The Twin-T Oscillator

The twin-T oscillator, shown in Fig. 10-6, is simply a modification of the twin-T "notch-in" filter discussed previously in Section 8-8. The oscillator, however, has a feedback path to the noninverting input. This positive feedback causes the circuit to go into oscillation at a frequency dictated by the component values of the twin-T network. See Eqs. (8-24) through (8-26). The output V_o is sinusoidal provided the feedback is not excessive. The amount of feedback, and consequently the output amplitude, is controlled by the 50 kΩ potentiometer. The parallel diodes alternately go into forward

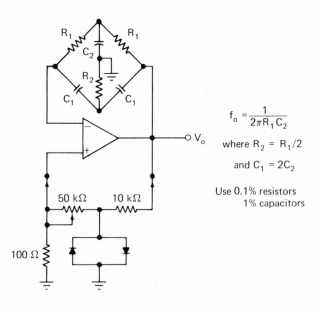

$$f_o = \frac{1}{2\pi R_1 C_2}$$

where $R_2 = R_1/2$

and $C_1 = 2C_2$

Use 0.1% resistors
1% capacitors

Fig. 10-6   Twin-T oscillator.

conduction with excessive feedback signals and thus limit the feedback amplitude to the noninverting input. To insure good frequency stability, polycarbonate or silver mica capacitors should be used. The twin-T's selectivity depends on how closely its components match. The use of 1% capacitors and 0.1% resistors minimizes the trimming needed to obtain good selectivity.

10-5   The Wien Bridge Oscillator

A Wien bridge oscillator is shown in Fig. 10-7. Its output is sinusoidal as is the twin-T oscillator's. In this case, the Op Amp's output is across a network consisting of R_1, C_1, R_2, and C_2 which essentially is the frequency-sensitive portion of a Wien bridge. The signal at point x of the Wien bridge

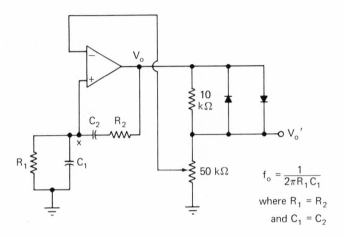

Fig. 10-7   The Wien bridge oscillator.

network, which is the input signal on the noninverting input, is in phase with the signal V_o at a particular frequency f_0. This frequency is

$$f_0 = \frac{1}{2\pi R_1 C_1} \tag{10-3}$$

if

$$R_1 = R_2$$

and

$$C_1 = C_2$$

The feedback signal at x leads V_o at frequencies below f_o and lags V_o at frequencies above f_0. Of course then, maximum in-phase feedback occurs at f_0, which therefore is the output frequency of the oscillator. Adjustment of the 50-kΩ potentiometer controls the amount of negative feedback to the inverting input and the amplitude of the output V_o'. As in the previous circuit, the diodes prevent excessive feedback amplitude. As with the twin-T oscillator, this circuit requires 1 % capacitors and 0.1 % resistors for reliable operation.

10-6   Variable-frequency Signal Generators

The circuit in Fig. 10-8 is an extension of the triangular-wave generator in Fig. 10-4. A broad range of output frequencies can be selected by the six-position switch shown. Each higher switch position increases the output frequency by ten. Of course, if we intend to use the high-frequency positions 5 and 6, we would select a suitable high-frequency Op Amp with a high slew rate. The 25-kΩ potentiometer is a fine frequency control. The 50-kΩ symmetry control enables us to vary the ratio of the time of each positive alternation to the time of each negative alternation of the waveforms V_o and

RANGE SELECTION

(1) 0.5 Hz to 5 Hz
(2) 5 Hz   to 50 Hz
(3) 50 Hz   to 500 Hz
(4) 500 Hz to 5 kHz
(5) 5 kHz   to 50 kHz
(6) Over 50 kHz

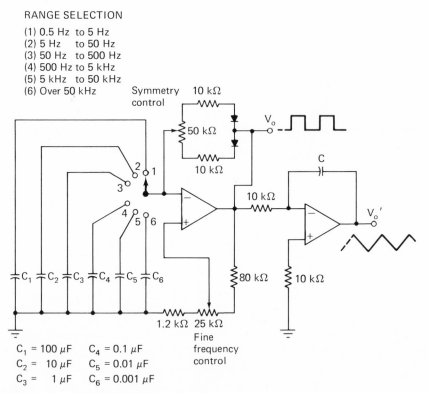

$C_1 = 100 \ \mu F$ $C_4 = 0.1 \ \mu F$
$C_2 = \ 10 \ \mu F$ $C_5 = 0.01 \ \mu F$
$C_3 = \ \ 1 \ \mu F$ $C_6 = 0.001 \ \mu F$

Fig. 10-8   Variable square-wave and triangular-wave generator.

V_o'. Thus the output V_o' can be changed from a triangular to a sawtooth waveform by the symmetry control. Capacitor C largely determines the amplitude of V_o'. Generally, the capacitor C must be larger with lower output frequencies. If C is too small, the output V_o' becomes clipped because the output Op Amp saturates on each alternation. If C is too large, the amplitude of V_o' is very small, especially at higher frequencies.

An easily tuned sine-wave and square-wave generator is shown in Fig. 10-9. Variations in the resistance R_1 vary the frequency, while resistance R_2 controls the amplitudes of the outputs. This circuit works on the principle of filtering and amplifying the fundamental frequency of a square wave.

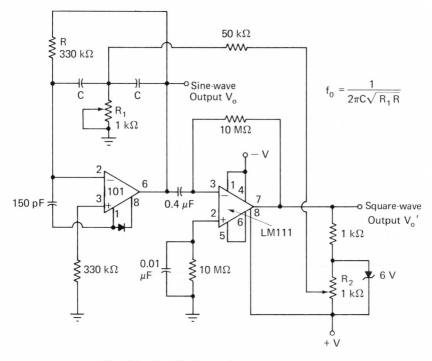

Fig. 10-9 Tunable sine- and square-wave generator.

As indicated, the first stage—the 101A Op Amp—has a sine-wave output V_o. This sine wave drives the second stage, the LM111. This second stage is a high-gain IC amplifier and works as a comparator, that is, the output of the LM111 is repetitively driven into saturation in positive and negative directions by successive alternations of the sine-wave input. The resulting square-wave output of the LM111 is fed back to the 101A via a filter network.

The filter essentially removes higher-order harmonics from the square wave and drives the first stage with its fundamental, resulting in a somewhat sinusoidal output. The zener clips the square-wave output V'_o and thus prevents its and the sine wave's amplitudes from drifting. As shown in Fig. 10-9, this circuit's output frequency is

$$f_0 = \frac{1}{2\pi C \sqrt{R_1 R}} \tag{10-4}$$

This circuit's components can be selected and adjusted to provide outputs from about 20 Hz to 20 kHz.

PROBLEMS

1. In the circuit of Fig. 10-1, if $R_a = 100 \, k\Omega$, $R_b = 10 \, k\Omega$, $R = 20 \, k\Omega$, and $C = 1 \, \mu F$, what is its approximate output frequency f_0?

2. If $R_a = 200 \, k\Omega$, $R_b = 20 \, k\Omega$, and $C = 1 \, \mu F$, what value of resistance R should we use in the circuit of Fig. 10-1 if we need approximately a 100-Hz output?

3. In the circuit described in Problem 1, about what maximum peak-to-peak value of output signal voltage can we expect if the Op Amp is a type 741 and the dc supply voltages are $+15$ V and -15 V?

4. If the Op Amp in the circuit of Fig. 10-3 is a type 741 operated with dc supply voltages of $+20$ V and -20 V, and if each zener has a zener voltage $V_z = 6.3$ V, what approximate peak-to-peak output signal voltage can we expect?

5. If C_1 is too small in the circuit of Fig. 10-4, what effect might it have on the output waveform V'_o?

6. If R_1 is too large in the circuit of Fig. 10-4, what effect will it have on the amplitude of the output waveform V'_o?

7. Sketch the approximate waveform of the output V'_o that we could expect from the circuit in Fig. 10-5 if $R_1 = 1 \, k\Omega$ and $R_2 = 20 \, k\Omega$?

8. Sketch the approximate waveform of the output V'_o that we could expect from the circuit in Fig. 10-5 if the diode becomes open?

9. With $R_1 = 160 \, k\Omega$ in the circuit of Fig. 10-6, what values of C_1, R_2, and C_2 should we use to obtain an output frequency of about 478 Hz?

10. If 0.05μF and 0.1μF capacitors are available, what values of R_1 and R_2 should we use in the circuit of Fig. 10-6 for a 318.4-Hz output?

11. Answer the question in Problem 9 for the Wien bridge instead of the twin-T oscillator.

12. Answer the question in Problem 10 for the circuit in Fig. 10-7 instead of Fig. 10-6.

CHAPTER **11**

DIGITAL
APPLICATIONS
OF OP AMPS

Digital applications of Op Amps are nonlinear and similar to some of the nonlinear circuits discussed in Chapter 9. The so-called *digital* circuits discussed here are the ones more likely to be associated with digital systems such as computers, calculators, and numerical controls. In typical digital applications, the outputs of the Op Amps are abruptly driven from one level to another.

11-1   The Schmitt Trigger

The Schmitt trigger circuit is similar to the comparator and zero-crossing detector circuits discussed previously. As its name implies, its purpose is to provide an output *trigger* voltage when its input signal reaches some predetermined value. In digital systems, rapidly changing voltages, called voltages with high de/dt, are used to trigger other circuits into their particular functions. The leading and trailing edges of square waveforms are often used as triggers. Therefore, circuits with square output voltages are sometimes called trigger circuits provided they themselves are stimulated by an input signal.

The output of an Op Amp type of Schmitt trigger abruptly swings from the Op Amp's positive saturation voltage $V_{o(\max)}$ to its negative saturation voltage $-V_{o(\max)}$, and vice versa. The Op Amp Schmitt trigger circuit and its transfer characteristics are shown in Fig. 11-1. As indicated by its transfer characteristic shown in Fig. 11-1b, the output is positively saturated as long as the input signal voltage V_s is less than the *upper threshold voltage V_{th}*. If V_s rises slightly above this threshold voltage V_{th}, the output swings to $-V_{o(\max)}$

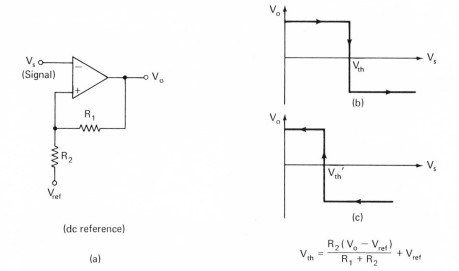

Fig. 11-1 (a) Schmitt trigger circuit, (b) transfer function for increasing V_s, and (c) transfer function for decreasing V_s.

and stays there until V_s drops below a *lower threshold voltage V_{th}'*. The threshold voltages are determined by the components R_1, R_2, and the dc reference voltage V_{ref}. These threshold voltages can be determined with the equation

$$V_{th} = \frac{R_2(V_o - V_{ref})}{R_1 + R_2} + V_{ref} \qquad (11\text{-}1)$$

where: V_o is the maximum positive output voltage when solving for the upper threshold, or

V_o is the maximum negative output voltage when solving for the lower threshold.

In practice, this circuit's peak-to-peak output is often limited by the use of back-to-back zeners across the output terminal and ground. The zener voltages are chosen so that the output swing is compatible with popular IC digital systems.

EXAMPLE

If in the circuit of Fig. 11-1, $R_1 = 10\,\text{k}\Omega$, $R_2 = 220\,\Omega$, $V_{ref} = 2\,\text{V}$, and the Op Amp saturates at $V_o = \pm 10\,\text{V}$, what are the upper and lower threshold voltages?

Answer. Since the Op Amp saturates positively at $+10$ V, the upper threshold voltage is

$$V_{th} = \frac{220 \ \Omega(10 \text{ V} - 2 \text{ V})}{10.22 \text{ k}\Omega} + 2 \text{ V} \cong 2.17 \text{ V}$$

This means that if the input voltage V_s rises slightly higher than 2.17 V, the Op Amp is driven into negative saturation, -10 V in this case.

With the output V_o as negative as -10 V, the lower threshold voltage is

$$V'_{th} = \frac{220 \ \Omega(-10 \text{ V} - 2 \text{ V})}{10.22 \text{ k}\Omega} + 2 \text{ V} \cong 1.74 \text{ V}$$

This means that if the input V_s drops slightly below 1.74 V, the output swings back to $+10$ V. Output vs input waveforms of this circuit are shown in Fig. 11-2.

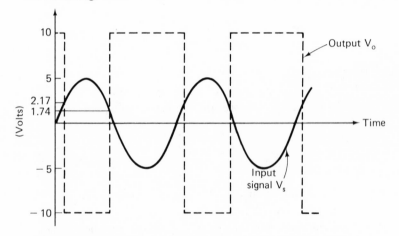

Fig. 11-2   Output V_o vs input V_s waveforms of the Schmitt trigger if the upper threshold voltage $V_{th} = 2.17$ V and the lower threshold $V'_{th} = 1.74$ V.

11-2   The Monostable (one-shot) Multivibrator

A monostable multivibrator provides an output voltage pulse of a specific time duration t after being excited by a pulse (trigger) of either long or short duration. The monostable circuit is commonly called a *one-shot.*

An Op Amp version of it is shown in Fig. 11-3. In the absence of an input $(V_i = 0\ \text{V})$, the Op Amp saturates positively and its output remains stable at $V_{o(max)}$.

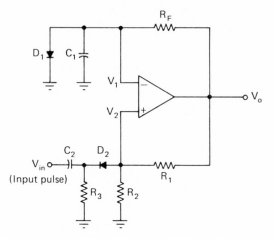

Fig. 11-3   An Op Amp monostable (one-shot) multivibrator.

At times when the Op Amp is positively saturated—in a stable state—the diode D_1 is forward biased, and the voltage across capacitor C_1, which is the voltage V_1 at the inverting input, is limited to a few tenths of a volt. At the same time, a portion of the output voltage V_o is fed back to the non-inverting input via the voltage divider R_1 and R_2.* Thus the voltage at the noninverting input is

$$V_2 = \frac{V_o R_2}{R_1 + R_2} \tag{11-2}$$

which is substantially larger than V_1 and holds the Op Amp in positive saturation.

Now if a negative going input pulse V_i is applied, whose peak-to-peak amplitude exceeds the voltage drop across R_2, the voltage V_2 at the non-inverting input is momentarily negative, and this drives the Op Amp into negative saturation. The voltage V_2 is only momentarily negative because of the short time constant of the coupling components $R_3 C_2$. With the output V_o now negative, the voltage V_2 is also negative, and this holds the Op Amp in negative saturation. This condition, however, is only temporary. The diode D_1, now reverse biased, permits capacitor C_1 to charge. As it charges, voltage V_1 increases negatively until it becomes slightly more negative than

*Assuming $R_3 >> R_2$

V_2, at which time it drives the Op Amp back into positive saturation. Thus, the output V_o remains negatively saturated only long enough for capacitor C_1 to charge to approximately the V_2 voltage. This time t is determined by the time constant of the C_1 and R_F and by the ratio of R_1/R_2. If the ratio $R_1/R_2 = 10$, then the time t of each negative pulse can be approximated with the equation:

$$t \cong 0.1(R_F C_1) \tag{11-3}$$

Waveforms associated with this monostable circuit are shown in Fig. 11-4.

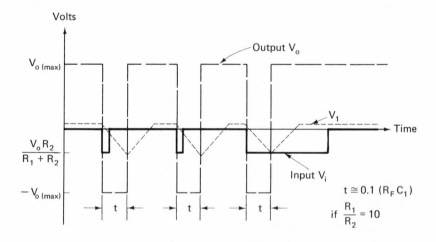

Fig. 11-4   Output V_o vs input V_i waveforms of the monostable multi-vibrator.

11-3   Digital-to-analog (D/A) Converters

Digital systems usually work with two levels of voltage referred to as HIGH and LOW signals or as logic 1 and logic 0. This two-level approach to performing computations and decisions is called a *binary system*, and values in such a system are expressed and processed in binary numbers. Some binary numbers along with their decimal equivalents are shown in Table 11-1. Such binary numbers are often read off of flip-flops followed by level amplifiers which can be viewed here as being equivalent to switches that are capable of providing *either* an output voltage V *or* 0 V as shown in Fig. 11-5.

Decimal Binary

	DCBA
0	0000
1	0001
2	0010
3	0011
4	0100
5	0101
6	0110
7	0111
8	1000
9	1001
10	1010
11	1011
12	1100
13	1101
14	1110
15	1111
16	10000

Table 11-1

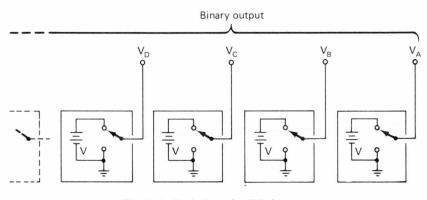

Fig. 11-5 Equivalent of a digital system.

When it is necessary to convert a binary output from a digital system to some equivalent analog voltage, a digital-to-analog (D/A) converter is used. An analog output should have 16 possible values if there are 16 combinations of digital voltages V_A through V_D. For example, since the binary number 0110 (decimal 6) is twice the value of the binary number 0011 (decimal 3), an analog equivalent voltage of 0110 is double the analog voltage representing 0011.

Binary outputs from digital systems can be converted to analog equivalent voltages by the use of either of the resistive networks of Fig. 11-6. The binary-weighted network in Fig. 11-6a, though requiring fewer resistors, needs a

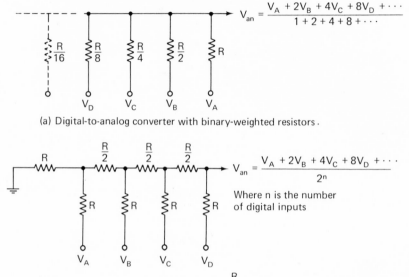

$$V_{an} = \frac{V_A + 2V_B + 4V_C + 8V_D + \cdots}{1 + 2 + 4 + 8 + \cdots}$$

(a) Digital-to-analog converter with binary-weighted resistors.

$$V_{an} = \frac{V_A + 2V_B + 4V_C + 8V_D + \cdots}{2^n}$$

Where n is the number of digital inputs

(b) Digital-to-analog converter with R and $\frac{R}{2}$ resistors.

Fig. 11-6   Resistive networks as digital-to-analog (D/A) converters.

variety of precision resistance values. Its analog output V_{an}, as a function of the two-level inputs, can be determined with the equation:

$$V_{an} = \frac{V_A + 2V_B + 4V_C + 8V_D + \cdots}{1 + 2 + 4 + 8 + \cdots} \tag{11-4}$$

If each of its digital inputs, V_A through V_D, is either 15 V or 0 V, the analog outputs V_{an} versus all possible combinations of inputs are shown in Table 11-2.

The D/A converter of Fig. 11-6b, using R and $R/2$ resistors, requires more resistors but only two sets of precision resistance values. Its output is

$$V_{an} = \frac{V_A + 2V_B + 4V_C + 8V_D + \cdots}{2^n} \tag{11-5}$$

where n is the number of digital inputs. If the levels of each of the inputs are 8 V and 0 V, all possible analog outputs of this four-input resistive network are shown in Table 11-3.

If the digital inputs to either network increase consecutively through larger decimal equivalents, the analog outputs are staircase waveforms as shown in Tables 11-2 and 11-3.

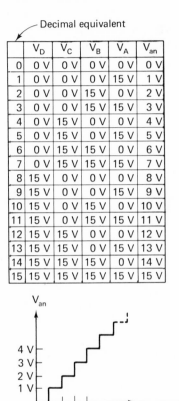

Decimal equivalent

	V_D	V_C	V_B	V_A	V_{an}
0	0 V	0 V	0 V	0 V	0 V
1	0 V	0 V	0 V	15 V	1 V
2	0 V	0 V	15 V	0 V	2 V
3	0 V	0 V	15 V	15 V	3 V
4	0 V	15 V	0 V	0 V	4 V
5	0 V	15 V	0 V	15 V	5 V
6	0 V	15 V	15 V	0 V	6 V
7	0 V	15 V	15 V	15 V	7 V
8	15 V	0 V	0 V	0 V	8 V
9	15 V	0 V	0 V	15 V	9 V
10	15 V	0 V	15 V	0 V	10 V
11	15 V	0 V	15 V	15 V	11 V
12	15 V	15 V	0 V	0 V	12 V
13	15 V	15 V	0 V	15 V	13 V
14	15 V	15 V	15 V	0 V	14 V
15	15 V	15 V	15 V	15 V	15 V

Decimal equivalent

	V_D	V_C	V_B	V_A	V_{an}
0	0 V	0 V	0 V	0 V	0 V
1	0 V	0 V	0 V	8 V	0.5 V
2	0 V	0 V	8 V	0 V	1 V
3	0 V	0 V	8 V	8 V	1.5 V
4	0 V	8 V	0 V	0 V	2 V
5	0 V	8 V	0 V	8 V	2.5 V
6	0 V	8 V	8 V	0 V	3 V
7	0 V	8 V	8 V	8 V	3.5 V
8	8 V	0 V	0 V	0 V	4 V
9	8 V	0 V	0 V	8 V	4.5 V
10	8 V	0 V	8 V	0 V	5 V
11	8 V	0 V	8 V	8 V	5.5 V
12	8 V	8 V	0 V	0 V	6 V
13	8 V	8 V	0 V	8 V	6.5 V
14	8 V	8 V	8 V	0 V	7 V
15	8 V	8 V	8 V	8 V	7.5 V

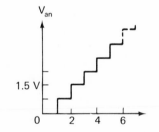

Table 11-2 Output vs inputs of the binary-weighted. resistive network where the digital logic levels are 0 V and 15 V.

Table 11-3 Output vs inputs of the R and $R/2$ resistive network where the digital logic levels are 0 V and 8 V.

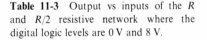

Both D/A converters of Fig. 11-6 must work into large-resistance loads; otherwise, the accuracy of their equations, (11-4) and (11-5), degenerates. Therefore, as shown in Fig. 11-7, an Op Amp is well suited to match a D/A resistive network to a low-resistance load and to provide gain as well. Placing an impedance-matching device, such as the Op Amp in this case, on the output of the resistive network is called *buffering* the output of the network.

11-4   Analog-to-digital (A/D) Converters

Often an analog voltage must be converted to a digital equivalent, such as in a digital voltmeter. In such cases, the principle of the previously discussed digital-to-analog D/A converter can be reversed to perform analog-to-digital A/D conversion. A system that converts analog voltages into digital equivalents is shown in Fig. 11-8.

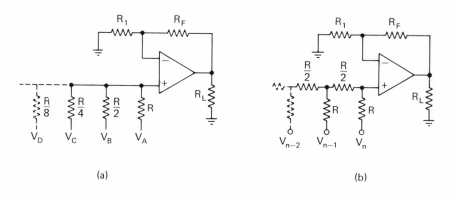

(a)   (b)

Fig. 11-7   Digital-to-analog (D/A) converters with buffered output and gain.

The up-down counter in the system of Fig. 11-8 has a digital output that increases with each clock pulse when its "count-up" line is HIGH and its "count-down" line is LOW. On the other hand, its digital output decreases with each clock pulse when its count-up line is LOW and its count-down line is HIGH.

The Op Amp U_2 works as a comparator. When its output is HIGH (positively saturated), the count-up line is also HIGH. Therefore, when U_2's output is LOW (negatively saturated), the count-up line is LOW too. Thus, depending on whether U_2's output is HIGH or LOW, the up-down counter counts digitally up or down, respectively. When the up-down counter is counting up, an upward staircase voltage appears at point *a*. When the counter is counting downward, a downward staircase is present at point *a*.

Since the Op Amp U_2 is operated in open loop, its output goes HIGH when its input 1 becomes *slightly* more negative than input 2. Conversely, its output goes LOW when its input 1 becomes *slightly* more positive than input 2.

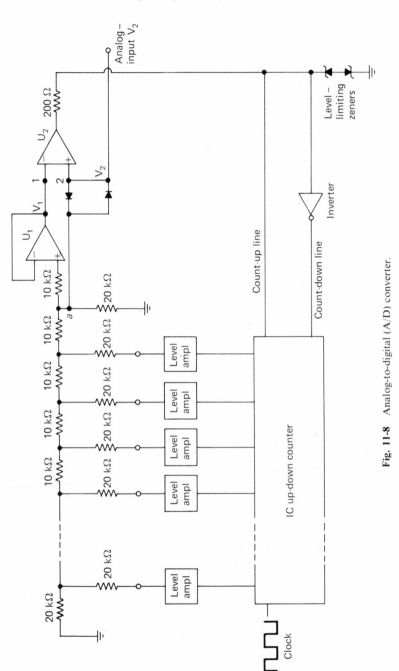

Fig. 11-8 Analog-to-digital (A/D) converter.

The first Op Amp U_1 is wired to work as a voltage follower and it buffers the D/A resistive network. Essentially then, the output voltage of the resistive network is applied to input 1 of U_2 and is thus compared with the analog input voltage V_2. This analog input V_2, of course, is the voltage to be digitized.

If the analog input voltage V_2 exceeds the resistive network's voltage V_1, the output of U_2 goes HIGH, and the up-down counter counts up, bringing the network's output voltage V_1 up, in steps, to the analog input V_2. On the other hand, if V_2 is less than or decreases below the voltage V_1, the output of U_2 goes LOW, and the up-down counter counts down, bringing voltage V_1 in line with V_2 again.

Generally then, this system has feedback which keeps the voltage output of the resistive network approximately equal to the analog input voltage V_2. In this way the output of the up-down counter is always a digital equivalent of the analog input V_2. Though not shown in Fig. 11-8, the counter's outputs can be used to drive a digital readout via an IC latch.

The level amplifiers, which sometimes can be Op Amps, are needed especially if the maximum analog input voltage V_2 exceeds the HIGH level output voltages of the up-down counter. The diodes prevent excessive differential inputs to U_2. The zeners are selected to clip the comparator's output to levels compatible with the up-down counter.

PROBLEMS

1. In the circuit of Fig. 11-1, if $R_1 = 40\ k\Omega$, $R_2 = 10\ k\Omega$, and $V_{ref} = -1$ V, what are the (a) upper and (b) lower threshold voltages? $V_{o(max)} = \pm 10$ V.

2. If $R_1 = 10\ k\Omega$, $R_2 = 1\ k\Omega$, and $V_{ref} = +2$ V in the circuit of Fig. 11-1, what are the (a) upper and (b) lower threshold voltages? $V_{o(max)} = \pm 12$ V.

3. In the circuit described in Problem 1, if the input signal V_s is a 100-Hz sine wave with a peak of 6 V, sketch the output waveform and indicate the approximate width of its positive and negative alternations in milliseconds.

4. In the circuit described in Problem 2, if V_s is a 400-Hz sine wave with a peak of 10 V, sketch the output waveform and indicate the approximate width of its positive and negative alternations in milliseconds.

5. In the circuit of Fig. 11-3, if $R_1 = 10\ k\Omega$, $R_2 = 1\ k\Omega$, $R_3 = 10\ k\Omega$, $R_F = 50\ k\Omega$, $C_1 = 0.1\ \mu F$, and $C_2 = 0.01\ \mu F$, what is the width of each negative pulse after the input is triggered with a negative pulse?

6. In the circuit described in Problem 5, with what resistance can we replace R_F to obtain a negative output pulse width of 1 ms?

7. What is the voltage V_o in the circuit of Fig. 11-9 if the switches 2^0, 2^1, 2^3, and 2^4 are in the 1, 1, 0, and 0 positions, respectively?

8. What is V_o in the circuit of Fig. 11-9 for each of the following sets of switch positions?

	2^0	2^1	2^2	2^3	V_o
(a)	1	0	0	0	
(b)	1	1	0	0	
(c)	0	0	1	0	

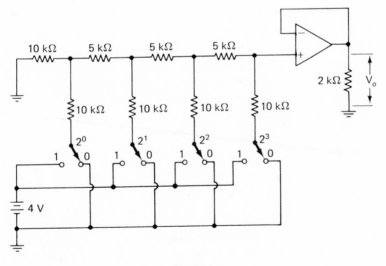

Fig. 11-9

CHAPTER 12

THE
CURRENT-DIFFERENCING
AMPLIFIER

The Op Amp's ability to provide an output voltage swing in both positive and negative directions is not always necessary or desirable. In such cases, the Op Amp's need of both positive and negative dc supply voltages is a costly inconvenience. The Op Amp can, of course, be wired to work with one dc supply as shown in Fig. 8-5, but this technique requires additional relatively costly components. A few manufacturers' answer to this problem is the *current-differencing** (CD) *amplifier*, also called the *Norton amplifier*. Furthermore, four CD amplifiers are available on a single 14-pin package which further reduces circuit size and cost. See Fig. 12-1. CD amplifiers are not one-to-one direct replacements for Op Amps. They do require special external wiring considerations and are not intended for dc voltage amplification.

12-1 The Basic Active Section of the CD Amplifier

The circuit of each CD amplifier is fundamentally as shown in Fig. 12-2. Q_1 is the active transistor of a common-emitter amplifier and Q_2 serves as its load resistance. The signal from the collector of Q_1 drives the base of Q_3 which operates as an emitter follower with its load being transistor Q_4. This arrangement is capable of gains over 60 dB, open loop. Transistors Q_2 and Q_4 are biased to behave as constant-current sources and therefore have very high dynamic resistances.

---

* Manufacturers' part numbers for current-differencing amplifiers are MC3401 (Motorola) and LM3900 (National Semiconductor).

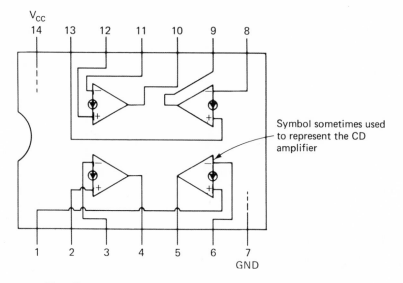

Fig. 12-1 The quad amplifier package contains four current-dfferencing amplifiers and a bias supply.

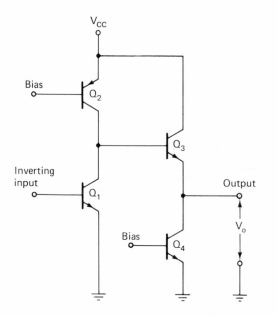

Fig. 12-2 Amplifying section of a current-differencing amplifier. Q_2 and Q_4 are biased to behave as high ac resistances (current sources).

Notice where the output voltage V_o is taken from in this circuit of Fig. 12-2. When the base of Q_3 is driven hard, causing this transistor to conduct very well, nearly all of the dc source voltage V_{CC} appears across Q_4 and is the output V_o. On the other hand, with little or no drive on Q_3, it is cut off and nearly all of the V_{CC} voltage appears across its collector and emitter terminals. In this case, the output voltage V_o, which is the voltage across Q_4, is about 0 V. Therefore, for linear applications and relatively large output signal capability, this amplifier should be biased so that the quiescent dc output voltage V_o is about half the V_{CC} voltage. As shown in Fig. 12-3, more clipping occurs with larger output signals if the output is not biased at $V_{CC}/2$.

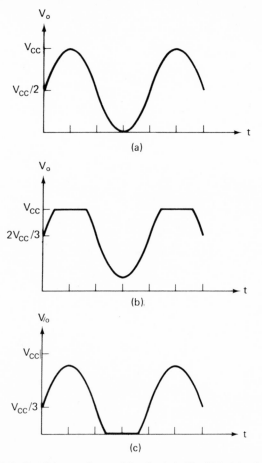

Fig. 12-3 Possible waveforms of the circuit in Fig. 12-2: (a) quiescent V_o at $V_{CC}/2$ gives maximum peak-to-peak output capability; (b) quiescent V_o higher than $V_{CC}/2$ causes clipping of positive peaks; (c) quiescent V_o lower than $V_{CC}/2$ causes clipping of negative peaks.

12-2   Feedback with the CD Amplifier

In the preceding chapters, we reduced the effective gain of an Op Amp by placing a feedback resistor across its output and inverting input. We can take the same approach to control the gain of the CD amplifier circuit. Use of a feedback resistor R_F, however, presents a problem. Since the CD amplifier, intended for linear operation, is biased so that its output is above ground potential, an average dc current I_F is forced through R_F as shown in Fig. 12-4. That is, the right end of R_F is above ground by the quiescent value of V_o, while its left end is above ground by the base-to-emitter voltage V_{BE} of Q_1, which typically is just a few tenths of a volt. This causes a dc potential difference across R_F that is $V_o - V_{BE}$; therefore the dc current through this feedback resistor is

$$I_F = \frac{V_o - V_{BE}}{R_F} \qquad (12\text{-}1)$$

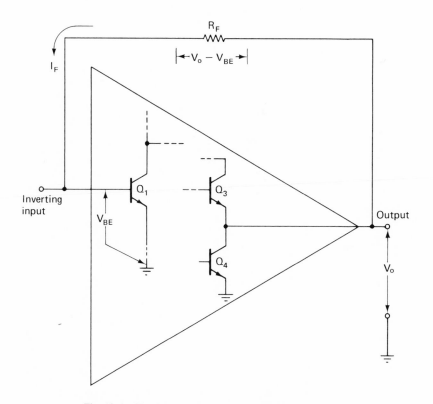

Fig. 12-4   The CD amplifier with a feedback resistor R_F.

If the dc supply voltage V_{CC} is considerably larger than V_{BE}, as it usually is, and if the quiescent value of V_o is to be half of the V_{CC} voltage, for good peak-to-peak output signal capability, Eq. (12-1) can be modified to

$$I_F \cong \frac{V_{CC}/2}{R_F} = \frac{V_{CC}}{2R_F} \qquad (12\text{-}2)$$

This feedback current I_F is much too large to properly bias the base of Q_1. In fact, if most of I_F is not diverted from the base of Q_1, the amplifier will saturate. Additional circuitry provides a shunt path for this dc feedback current I_F as shown in Fig. 12-5.

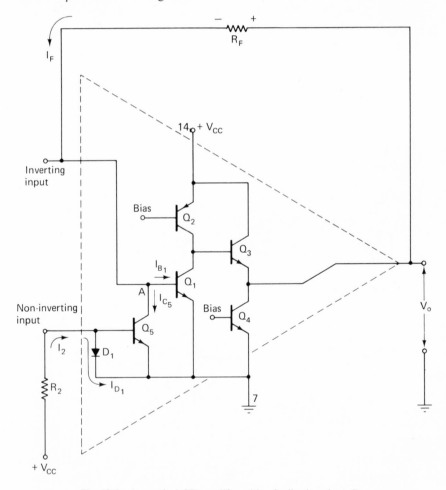

Fig. 12-5   A practical CD amplifier with a feedback resistor R_F.

In this practical circuit of a current-differencing amplifier (Fig. 12-5), the transistor Q_5, with proper bias, shunts most of the dc feedback current I_F to ground. Q_5 diverts I_F and prevents excessive drive on the base of Q_1, and this prevents the amplifier's output from saturating. In fact, by controlling the amount of current diverted by Q_5, we select the quiescent value of output voltage V_o.

As shown in Fig. 12-5, current I_F flows to point A. At this point, it divides into currents I_{B_1} and I_{C_5}, where I_{B_1}* is the base current of Q_1 and I_{C_5} is the collector current of Q_5. Typically, I_F is much larger than the base bias current I_{B_1} required to adjust the output voltage V_o to a useful quiescent value, such as to $V_{CC}/2$. Therefore, most of I_F *should* become I_{C_5}, that is, for linear operation

$$I_F \cong I_{C_5} \tag{12-3a}$$

If I_{C_5} is too large or too small, the quiescent value of V_o hangs up at about the V_{CC} voltage or drops to about 0 V, causing severe distortion (clipping) of the output signal waveform.

The value of I_{C_5} is controlled by the amount of forward bias on the base-emitter junction of Q_5, which in turn is determined by the forward drop across the diode D_1. This diode D_1 and transistor Q_5 have characteristics such that Q_5 conducts an I_{C_5} that is about equal to the diode's current I_{D_1}:

$$I_{C_5} \cong I_{D_1} \tag{12-3b}$$

The amount of I_{D_1} through D_1 is determined by the values of the externally wired bias resistor R_2 and the dc source V_{CC}. Since the top of R_2 is above ground by the base-emitter voltage V_{BE} of Q_5, the current through R_2 is, by Ohm's law,

$$I_2 = \frac{V_{CC} - V_{BE}}{R_2} \tag{12-4a}$$

Of course, if V_{CC} is much greater than V_{BE}, the Eq. (12-4a) can be more simply stated as

$$I_2 \cong \frac{V_{CC}}{R_2} \tag{12-4b}$$

---

* I_{B_1} is the difference in currents I_F and I_{C_5}, and since it is the input to Q_1, the circuit is called a current-differencing amplifier.

Most of I_2 flows through D_1. Combining this fact with Eqs. (12-3a) and (12-3b), we can show that

$$I_2 \cong I_F \qquad (12-5)$$

for linear operation. In fact, the noninverting input's dc current I_2 is called a *mirror current* when it reflects (is equal to) the dc feedback current I_F.

It is a simple matter to select components that will establish a mirror current and linear operation with good peak-to-peak output signal capability. For example, since I_2 should approximately equal I_F, we can set the right sides of Eqs. (12-2) and (12-4b) equal to each other; therefore

$$\frac{V_{CC}}{2R_F} \cong \frac{V_{CC}}{R_2}$$

which simplifies to

$$R_2 \cong 2R_F \qquad (12-6)$$

This last equation is a handy one for building inverting-mode and non-inverting-mode amplifiers, as we will see; but it is valid only if we tie one end of R_2 to the V_{CC} source voltage.

12-3 The Inverting Current-differencing Amplifier

A simple inverting circuit using a CD amplifier is shown in Fig. 12-6. It can be used to amplify ac but not dc signals because of the input coupling capacitor C_1. This capacitor is needed to block the dc feedback current I_F.

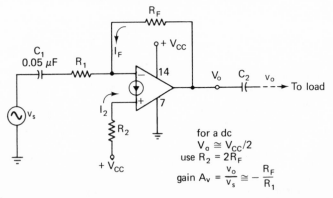

Fig. 12-6   The CD amplifier connected to work in an inverting mode.

Without C_1, a large part of current I_F would flow through R_1 and the internal resistance of the signal source V_s. This would probably cause a drifting current into the inverting input, making it difficult or impossible to select a proper mirror current and operating point. As shown in Fig. 12-6 and according to Eq. (12-6), resistor R_2 is selected to be twice the value of R_F, and one end of R_2 is tied to the V_{CC} voltage. This assures us of a good operating point, that is, a quiescent output V_o that is about half of the V_{CC} voltage.

From the ac signal's point of view, this circuit can be analyzed as the conventional Op Amps were in preceding chapters, resulting in a similar gain equation. At frequencies that see the reactances of the coupling capacitors C_1 and C_2 as negligible (see Appendix A3), this inverting circuit's gain is

$$A_v = \frac{V_o}{V_s} \cong -\frac{R_F}{R_1} \qquad (12\text{-}7)$$

The feedback resistor R_F is selected so that the feedback current is about equal to the manufacturer's recommended mirror current. Once the required R_F is established, the value of R_1 is selected to provide the desired stage gain. Then R_2 is selected to inject a mirror current into the noninverting input that is equal to the feedback current I_F.

EXAMPLE 12-1

Select resistance values of R_F, R_1, and R_2 for the circuit in Fig. 12-6, using a CD amplifier, to provide a voltage gain of 100. The CD amplifier manufacturer's recommended mirror current is $10\,\mu A$ and the dc source $V_{CC} = +20$ V.

Answer. Since $I_F = V_{CC}/2R_F$ according to Eq. (12-2), and since it is to be equal to the recommended mirror current, we can solve for R_F and show that

$$R_F = \frac{V_{CC}}{2I_F} = \frac{20\text{ V}}{2(10\,\mu A)} = 1\text{ M}\Omega$$

Therefore, for a gain of 100, the input resistor R_1 must be smaller by the factor 100. By rearranging Eq. (12-7), we can show that

$$R_1 \cong \frac{R_F}{-A_v} = \frac{1\text{ M}\Omega}{100} = 10\text{ k}\Omega$$

Then for the required mirror current I_2,

$$R_2 \cong 2R_F = 2(1\text{ M}\Omega) = 2\text{ M}\Omega \qquad (12\text{-}6)$$

The CD amplifier can also be operated with a single negative dc supply as shown in Fig. 12-7. In this case, the ground pin of the package is connected to the $-V_{CC}$ source, and the $+V_{CC}$ pin of the package is grounded or connected to the common of the $-V_{CC}$ source. In fact, all previously grounded points of the IC are placed at $-V_{CC}$ potential, and all points previously at $+V_{CC}$ are grounded. The external component selection for this circuit is made in the same way as for the previous circuit (Fig. 12-6).

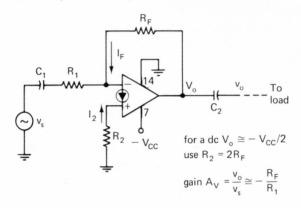

for a dc $V_o \cong -V_{CC}/2$
use $R_2 = 2R_F$

gain $A_V = \dfrac{V_o}{V_s} \cong -\dfrac{R_F}{R_1}$

Fig. 12-7   A CD amplifier connected to work with a negative supply.

12-4   The Noninverting Current-differencing Amplifier

The circuit in Fig. 12-8 is connected to work as a noninverting CD amplifier using a single positive dc source voltage $+V_{CC}$. The bias- and gain-determining components for this noninverting circuit are selected exactly as they are for the inverting types. Of course, the negative sign in the gain equation is not used because of the in-phase relationship of the output and input signal voltages in this case.

EXAMPLE 12-2

Select values of R_F, R_1, and R_2 for the noninverting circuit in Fig. 12-8, where the voltage gain is to be about 200 and the recommended mirror current is 50 μA. The dc supply $V_{CC} = 15$ V.

Answer.   Since the feedback and mirror currents are to be about equal for linear operation, we first solve for R_F by rearranging Eq. (12-2).

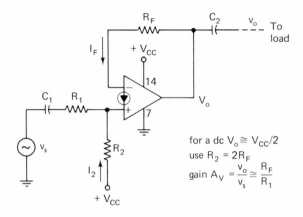

for a dc $V_o \cong V_{CC}/2$
use $R_2 = 2R_F$
gain $A_V = \dfrac{v_o}{v_s} \cong \dfrac{R_F}{R_1}$

Fig. 12-8 The CD amplifier connected to work as a noninverting amplifier.

Thus

$$R_F = \frac{V_{CC}}{2I_F} = \frac{15\,\text{V}}{2(50\,\mu\text{A})} = 150\,\text{k}\Omega$$

To obtain a gain $A_v = 200$, R_1 must be

$$R_1 \cong \frac{R_F}{A_v} = \frac{150\,\text{k}\Omega}{200} = 750\,\Omega \tag{12-7}$$

and by Eq. (12-6),

$$R_2 \cong 2R_F = 2(150\,\text{k}\Omega) = 300\,\text{k}\Omega$$

12-5 The CD Amplifier in Differential Mode

CD amplifiers can be wired to work in differential mode as shown in Fig. 12-9. With proper selection of components, this circuit has good common-mode rejection (*CMRR*), which means that ac or rf noise voltages induced into its input leads do not find their way to the load at the output. For a dc quiescent output $V_o = V_{CC}/2$, which gives us a good peak-to-peak output signal capability,

$$R_b + R_c = 2R_F \tag{12-8}$$

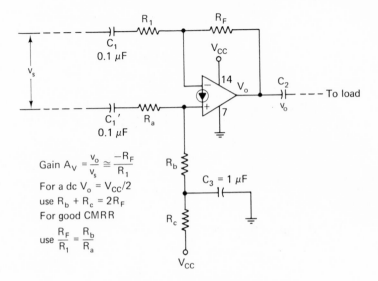

Fig. 12-9 The CD amplifier connected to work in a differential mode.

Similar to previous circuits, this circuit's voltage gain is

$$A_v = \frac{v_o}{v_s} \cong - \frac{R_F}{R_1} \tag{12-7}$$

and for good *CMRR*

$$R_b = R_F \tag{12-9a}$$

and

$$R_a = R_1 \tag{12-9b}$$

The capacitor C_3 places the bottom of resistor R_b at ac ground potential and is needed for good *CMRR*. As in the inverting and noninverting circuits discussed in the previous sections, the value of R_F is dictated by the value of the dc supply V_{CC} and the recommended mirror current.

EXAMPLE 12-3

Select resistor values for the circuit in Fig. 12-9 for good *CMRR* and a voltage gain of 120. The dc supply $V_{CC} = 24$ V and the recommended mirror current is 100 μA.

Answer. First, by Eqs. (12-2) and (12-9a)

$$R_F = R_b = \frac{V_{CC}}{2I_F} = \frac{24 \text{ V}}{2(100 \text{ }\mu\text{A})} = 120 \text{ k}\Omega$$

Now we find that

$$R_1 = R_a \cong \frac{R_F}{A_v} = \frac{120 \text{ k}\Omega}{120} = 1 \text{ k}\Omega \quad (12\text{-}7) \text{ and } (12\text{-}9b)$$

therefore

$$R_b + R_c = 2R_F = 2(120 \text{ k}\Omega) = 240 \text{ k}\Omega \qquad (12\text{-}8)$$

and finally

$$R_c = 240 \text{ k}\Omega - R_b = 120 \text{ k}\Omega$$

In addition to the linear circuits just discussed, the CD amplifier has countless nonlinear applications too. A few of these are discussed in the following sections.

12-6   The CD Amplifier as a Comparator

Comparators are used extensively in digital-to-analog converters and in analog-to-digital converters. Their function is to provide fast changes in output voltage levels when their input voltages V_i change through a reference voltage. Two CD comparators are shown in Fig. 12-10. In the circuit shown in Fig. 12-10a, the output $V_o \cong 0$ V as long as the input voltage V_i is more positive than the reference voltage V_{ref}. If V_i changes to a value less positive

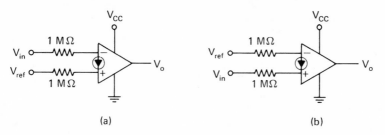

Fig. 12-10 CD amplifiers used as comparators: (a) $V_o \cong 0$ when $V_i > V_{\text{ref}}$, and $V_o \cong V_{CC}$ when $V_i < V_{\text{ref}}$; (b) $V_o \cong 0$ when $V_i < V_{\text{ref}}$, and $V_o \cong V_{CC}$ when $V_i > V_{\text{ref}}$.

than V_{ref}, the output V_o quickly swings to about the V_{CC} voltage. In the circuit shown in Fig. 12-10b, the output $V_o = 0$ V as long as the input V_i is less positive than the reference V_{ref}. When V_i swings to a value more positive than V_{ref}, the output V_o quickly switches to approximately the V_{CC} voltage.

The CD amplifier makes an efficient comparator because it requires only two current-limiting resistors in its input leads. No feedback resistor is used because a high gain, such as the open-loop gain, is desirable. The high gain enables us to drive the output V_o between saturation and cutoff with very small changes in the input currents. With nonlinear applications such as these, no thought need be given to the mirror current. As mentioned before, the mirror current is important in linear applications because it is used to establish an operating point of the output V_o that is *between* V_{CC} and 0 V.

12-7   The CD Amplifier as a Square-wave Generator

The circuit in Fig. 12-11 uses a CD amplifier and works as a square-wave generator. With the fixed resistance values shown, components R_F and C determine the output frequency which can be approximated with the equation

$$f \cong \frac{1}{1.4R_F C} \tag{12-10}$$

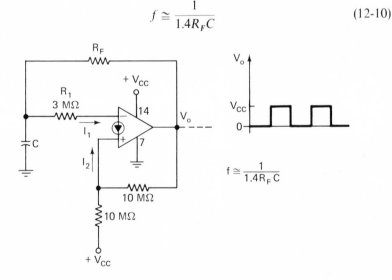

Fig. 12-11 A CD amplifier connected to work as a square-wave generator.

When the source V_{CC} is applied to this circuit, the output rises to about the V_{CC} voltage. It stays there until the capacitor C *charges* to a voltage large enough to drive a current I_1 into the inverting input that is slightly larger than the noninverting input's current I_2. Current I_2 at this time is the sum of the currents through each of the 10-MΩ resistors. When I_1 becomes larger than I_2, the output V_o quickly swings down to about 0 V and this simultaneously reduces I_2 to about half of its previous value. V_o remains at 0 V until the capacitor C now *discharges* to the point where I_1 is a value slightly less than I_2. When this happens, V_o quickly swings back up to the V_{CC} voltage to start another cycle.

12-8   The CD Amplifier as an **AND** Gate

A circuit that works as an **AND** gate has two or more inputs and an output. Its output cannot be a logic "1" (a positive voltage) until all of its inputs are logic "1" (positive). If any one of its inputs is logic "0" (about 0 V), the output is "0." A CD amplifier connected to work as an **AND** gate is shown in Fig. 12-12. Its output X is "1" only when its inputs A *and* B *and* C are "1." In equation form, this is shown as

$$X = A \times B \times C$$

or simply

$$X = ABC$$

Normally in the circuit of Fig. 12-12, each of the inputs has an applied voltage that is either the V_{CC} voltage or about 0 V. Only when all three inputs

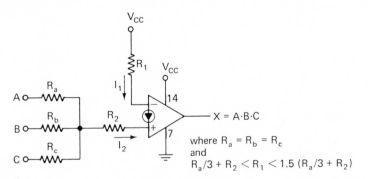

Fig. 12-12   A CD amplifier connected to work as an **AND** gate.

are at the V_{CC} voltage is the current I_2 greater than the current I_1, and this causes an output $V_o \cong V_{CC}$. If any one or more inputs is 0 V, the current I_2 is less than I_1, and this causes an output $V_o \cong 0$ V.

12-9   The CD Amplifier as an **OR** Gate

An **OR** gate has an output "1" if any one of its inputs is "1." Thus an **OR** gate with three inputs A, B, and C and output X has a logic "1" at the output if input A *or* input B *or* input C is "1." In equation form, this is shown as

$$X = A + B + C$$

The circuit in Fig. 12-13 shows a CD amplifier connected to work as an **OR** gate. The voltages applied to the inputs are typically either V_{CC} or 0 V which represent logic levels "1" or "0," respectively. The V_{CC} voltage at any one of the inputs will force a current I_2 that is larger than I_1 into the noninverting input, forcing V_o up to the V_{CC} voltage. Of course, with V_{CC} at more than one input, the current I_2 is much larger than I_1, and this drives V_o up to the V_{CC} voltage all the harder. Only when all inputs are at about 0 V is the current I_2 less than I_1, causing an output $V_o \cong 0$ V.

As with the conventional Op Amp, the possible applications of the CD amplifier are practically limitless. A number of them are given in the circuits section of this chapter. The symbol for the CD amplifier shown in the beginning sections of this chapter is not always used in practice. As shown in the circuits section, the Op Amp symbol is used to represent the CD amplifier, but then the part number is clearly identified in each circuit to avoid confusion.

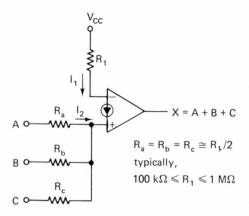

Fig. 12-13   A CD amplifier connected to work as an **OR** gate.

12-10   Collection of Current-differencing (Norton) Amplifier Circuits

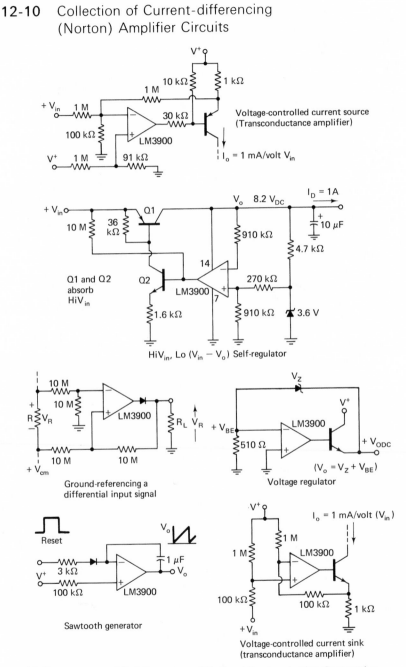

All diagrams on pp. 249–254 courtesy of National Semiconduction Corporation.

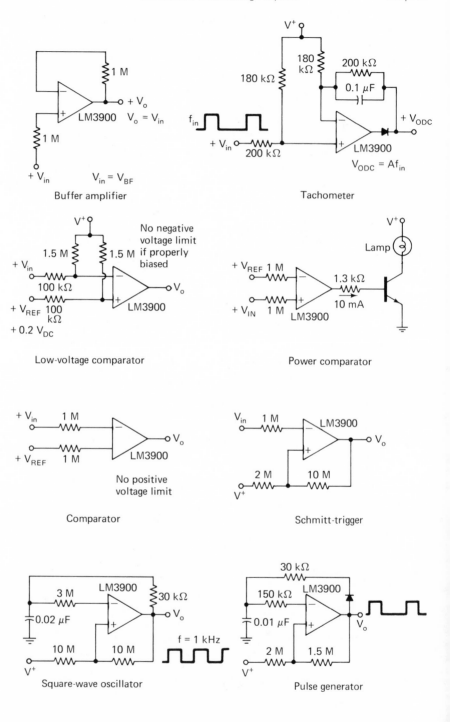

Buffer amplifier

Tachometer

Low-voltage comparator

Power comparator

Comparator

Schmitt-trigger

Square-wave oscillator

Pulse generator

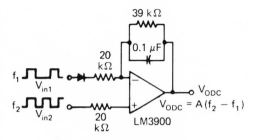

Frequency differencing tachometer

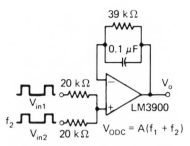

Frequency averaging tachometer

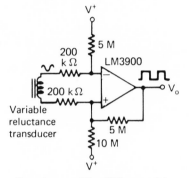

Squaring amplifier (W/hysteresis)

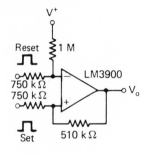

Bi-stable multivibrator

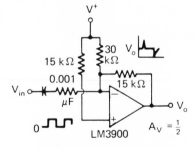

Differentiator (common-mode biasing keeps input at $+V_{BE}$)

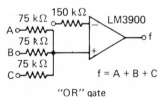

"OR" gate

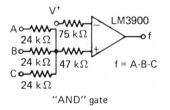

"AND" gate

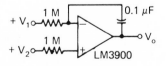

Difference integrator

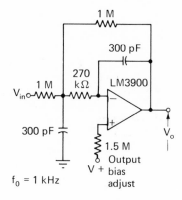

$f_0 = 1$ kHz

Low-pass active filter

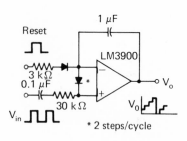

* 2 steps/cycle

Staircase generator

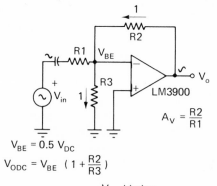

$$A_V = \frac{R2}{R1}$$

$$V_{BE} = 0.5 \, V_{DC}$$

$$V_{ODC} = V_{BE} \left(1 + \frac{R2}{R3} \right)$$

V_{BE} biasing

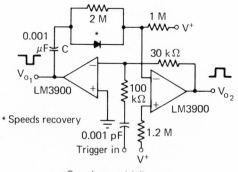

* Speeds recovery

One-shot multivibrator

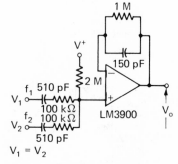

$V_1 = V_2$

Low-frequency mixer

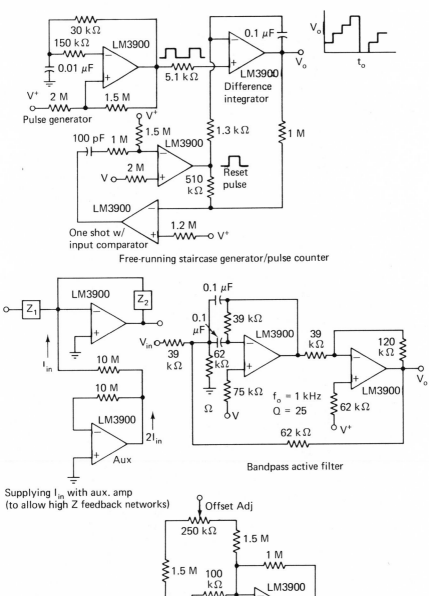

Free-running staircase generator/pulse counter

Supplying I_{in} with aux. amp
(to allow high Z feedback networks)

Bandpass active filter

Non-inverting DC gain

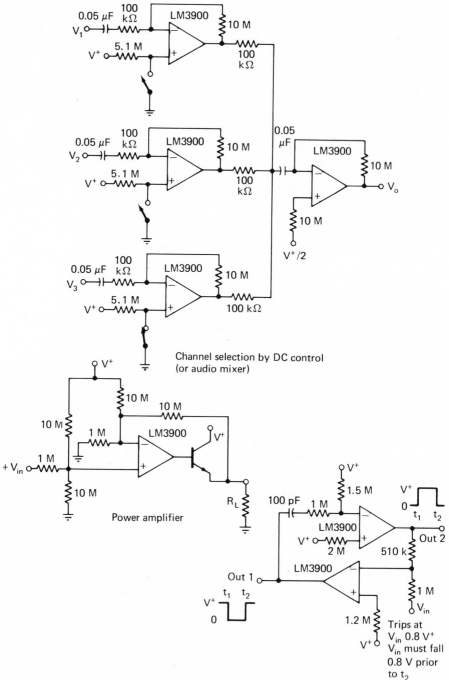

Channel selection by DC control (or audio mixer)

Power amplifier

One-shot w/DC input comparator

REVIEW QUESTIONS AND PROBLEMS

1. Name three advantages that CD amplifiers have compared to conventional Op Amps.

2. Name three disadvantages that CD amplifiers have compared to conventional Op Amps.

3. What type of CD amplifier applications require a mirror current?

4. How does the quiescent dc output of the CD amplifier compare to the dc source voltage if the dc feedback current is equal to the mirror current?

5. What might happen to the output signal waveform of the CD amplifier if its dc feedback current to the inverting input is not equal to the current in the noninverting input?

6. Why does a dc feedback current exist when the CD amplifier is biased for linear operation?

7. If we need comparators whose outputs are to swing between positive and zero volts, which is probably more suitable, the CD amplifier or the Op Amp? Why?

8. If we need a low-frequency square-wave generator whose output is to swing between positive and negative voltages, which is likely to be more suitable, the CD amplifier or the Op Amp? Why?

9. Can the differential-mode circuit in Fig. 12-9 be used to amplify signals from a thermocouple bridge that is monitoring the temperature of a large furnace? Why?

10. What circuit performance will suffer in the circuit of Fig. 12-9 if the capacitor C_3 becomes open?

11. In the circuit of Fig. 12-6, what is likely to happen to the output signal waveform if the signal source v_s is directly coupled into R_1 instead of through the coupling capacitor C_1?

GLOSSARY
OF TERMS

Amplification, Differential (A_d)   The ratio of the output voltage to the differential input voltage of a differential amplifier.

Amplification, Voltage (A_v)   The closed-loop voltage gain of an Op Amp which is its effective gain with negative feedback; also called the closed-loop gain.

Amplification, Voltage Open Loop (A_{VOL})   The ratio of the output voltage to the differential input voltage with no feedback also called the open-loop gain.

Average Temperature Coefficient of Input Offset Current   The operating temperature range divided into the resulting change in the input offset current.

Average Temperature Coefficient of Input Offset Voltage   The operating temperature range divided into the resulting change in the input offset voltage.

Bandwidth (BW)   The frequency range in which the Op Amp's open-loop gain does not drop more than 0.707 or 3 dB of its dc value.

Beta (β)   The ratio of the bipolar transistor's collector current to its base current; also called h_{fe}.

Bootstrapping   Technique of using feedback to raise an amplifier's input impedance.

Buffering   Use of an isolating circuit that prevents variations in load resistance from affecting the source of signal driving it.

Channel Separation   The ratio of the output signal voltage of a driven amplifier to the output signal voltage of an adjacent undriven amplifier; usually expressed in decibels.

Chip A small piece of semiconductor in or on which an integrated circuit is built.

Chopper-Stabilized Op Amp A high-performance Op Amp having a very small input bias current and low drift.

Common-Mode Gain The ratio of the common-mode output voltage to the common-mode input voltage; typically is much less than unity for differential and operational amplifiers.

Common-Mode Input Resistance The resistance with respect to ground or a common point looking into both inputs tied together.

Common-Mode Input Voltage Swing The peak value of common-mode input voltage that can be applied and still maintain linear operation.

Common-Mode Output Voltage The output voltage resulting from a voltage applied to both inputs simultaneously.

Common-Mode Rejection Ratio ($CMRR$) The ratio of the closed-loop gain to the common-mode gain; also, the ratio of the change in common-mode input voltage to the resulting change in input offset voltage; usually expressed in decibels.

Compensation Use of externally wired components to stabilize Op Amps that are not internally compensated.

Darlington Pair Two low-leakage bipolar transistors wired so that their total beta is the product of their individual betas.

Differential Input Resistance Resistance seen looking into the Op Amp's inputs under open-loop condition.

Differential Gain (A_b) See Amplification.

Drift Changes in parameters with changes in temperature, supply voltage, or time.

Drive Application of signal to the input of an amplifier.

Effective Input Resistance (R_i') Small-signal ac input resistance seen looking into the appropriate input with the other input grounded or common under closed-loop condition.

Effective Output Resistance (R_o') Small-signal ac resistance seen looking back into the Op Amp's output terminal under closed-loop condition.

Feedback   Use of circuitry that applies a portion of the Op Amp's output signal back to its inverting input.

FET   Field Effect Transistor.

Gain-Bandwidth Product   The product of a compensated Op Amp's closed-loop gain and its bandwidth with that gain.

Hybrid Op Amp   Op Amp containing ICs and discrete components.

IGFET   Insulated Gate Field Effect Transistor; also called a MOSFET.

Input Bias Current (I_B)   The average of the two dc input bias currents measured while both inputs are grounded or connected to a common point.

Input Bias Current Drift   The change in input bias current with change in temperature, supply voltage, or time.

Input Offset Current (I_{io})   The difference in the two dc input bias currents measured while both inputs are grounded or connected to a common point.

Input Offset Voltage (V_{io})   The voltage that must be applied across the inputs to force the output to zero volts.

Input Resistance (R_i)   The resistance seen looking into either input with the other grounded or common under open-loop condition.

Input Voltage Range   The range of voltage that can be applied to either input over which the Op Amp is linear.

Internal Power Dissipation   The power required to operate the amplifier with an open load and no input signal.

Large-Signal Voltage Gain   The open-loop gain A_{VOL} measured while driving the Op Amp so that its output voltage swing is relatively large but not clipped.

Latch-Up   A condition where an Op Amp's output stays in saturation after the input voltage causing it is gone.

Loop Gain   The ratio of the open-loop gain to the closed-loop gain.

Monolithic ICs   An integrated circuit whose components are formed on or within a single piece of semiconductor called a substrate.

MOSFET Metal-Oxide Semiconductor Field Effect Transistor; see IGFET.

Noise Figure (NF)   The ratio of an Op Amp's input signal-to-noise ratio to its output signal-to-noise ratio expressed in decibels.

Open-Loop Gain (A_{VOL})   The ratio of the output voltage to the differential input voltage with no feedback.

Output Protection   Use of a resistor in series with the output terminal to prevent excessive output currents.

Output Resistance (R_o)   Resistance seen looking back into the Op Amp's output terminal under open-loop condition.

Output Short-Circuit Current   The output current with output shorted to ground or common or connected to either dc supply.

Output Voltage Swing   The maximum output voltage swing that can be obtained without clipping.

Overshoot   The output voltage swing beyond its final quiescent value in response to a step input voltage.

Power Supply Current   The current an Op Amp draws from the power supply with the load open and no signal applied.

Power Supply Rejection Ratio ($PSRR$)   The ratio of the change in the power supply voltage to the resulting change in input offset voltage; usually expressed in decibels.

Power Supply Sensitivity   The ratio of the change in the input offset voltage to the change in power supply voltage causing it. The ratio of the change in a specified parameter to the change in the supply voltage causing it.

Rate of Closure   The rate at which the open-loop gain vs frequency curve is decreasing (rolling off) as it intersects the closed-loop vs frequency curve; usually expressed in decibels per octave or decibels per decade.

Ringing   The oscillations of the output voltage about the final quiescent value in response to a step input voltage.

Rise Time   The time required for an output voltage step to change from 10% to 90% of its final value.

Roll-Off   The decrease in amplifier gain at higher frequencies.

Saturation   An Op Amp driven to its maximum positive or its maximum negative output voltage; largely determined by the power supply voltages and load resistance.

Settling Time   The time between the instant a step input voltage is applied to the instant the output settles to a specified percentage of the final quiescent value.

Slew Rate   The maximum rate of change of the output voltage under large-signal conditions.

Standby Current Drain   The part of the output current of a power supply that does not contribute to the load current.

Tailored Response   Term used to describe the response of Op Amps made to be externally compensated and meet special requirements not available with internally compensated types.

Temperature Stability   The change in output voltage or voltage gain over a specified temperature range.

Transfer Function   The output voltage vs input voltage curve.

Transient Response   The output voltage response to a step input voltage under closed-loop and small-signal conditions.

Unity Gain Bandwidth   The bandwidth when the Op Amp's open-loop gain is down to unity.

Zener Knee Current   The minimum zener current required to keep the zener in zener (avalanche) conduction.

Zener Test Current   The zener current with which the specified zener voltage and zener resistance were measured.

SELECTION GUIDE
FOR GENERAL PURPOSE
OPERATIONAL AMPLIFIERS

Selection Guide for Commercial Operational Amplifiers

		General Purpose								
					Tailored Response					
		μA702*	μA709	μA739	μA748	μA749	μA777	201	201A	301A
		DC Wide Band		Dual Low Noise		Dual				
Input Offset Voltage	Max (mV)	5.0	7.5	6.0	6.0	6.0	7.5	7.5	2.0	7.5
Input Offset Current	Max (nA)	2000	500	1000	200	500	50	200	10	50
Input Bias Current	Max (nA)	7500	1500	2000	500	1000	250	500	75	250
Voltage Gain	Min (V/mV)	2.0	15	6.5	20	15	25	20	25	25
Operating Supply Voltage Range	Min (V)	+6.0, −3.0	±9.0	±4.0	±5.0	±4.0	±5.0	±5.0	±5.0	±5.0
	Max (V)	+14, −7.0	±18	±18	±18	±18	±20	±20	±20	±20
Unity Gain Bandwidth	Typ (MHz)	30	1.0	10	1.0	10	1.0	1.0	1.0	1.0
Slew Rate $A_{CL} = 1$	Typ (V/μs)	3.5	0.3	1.0	0.5	1.5	0.5	0.5	0.5	0.5
$A_{CL} = -1$		3.5	0.3	2.5	6.0	2.5	6.0	6.0	6.0	6.0
$A_{CL} = 10$		5.0	3.0	8.0	2.0	8.0	2.0	2.0	2.0	2.0
Input Voltage Range	Max (V)	+1.5, −6.0	±10	±15	±15	±15	±15	±15	±15	±15
Differential Input Voltage	Max (V)	±5.0	±5.0	±5.0	±30	±5.0	±30	±30	±30	±30
Input Offset Voltage Drift	Typ (μV/°C)	10	10	4.0		7.0	3.0	7.0	3.0	6.0
Offset Adjust					X		X	X	X	X
Output Short Circuit Protection					X		X	X	X	X
Compensated										
Dual				X		X				

Selection Guide for Military Operational Amplifiers

		General Purpose							
					Tailored Response				
		μ702*	μA709A	μA709	μA748	μA749	μA777	101	101A
		DC Wide Band			Dual				
Input Offset Voltage	Max (mV)	2.0	2.0	5.0	5.0	3.0	2.0	5.0	2.0
Input Offset Current	Max (nA)	500	50	200	200	400	10	200	10
Input Bias Current	Max (nA)	5000	200	500	500	750	75	500	75
Voltage Gain	Min (V/mV)	2.5	25	25	50	25	50	50	50
Operating Supply Voltage Range	Min (V)	+6.0, −3.0	±9.0	±9.0	±5.0	±4.0	±5.0	±5.0	±5.0
	Max (V)	+14, −7.0	±18	±18	±22	±18	±20	±20	± 20
Unity Gain Bandwidth	Typ (MHz)	30	5.0	5.0	1.0	10	1.0	1.0	1.0
Slew Rate $A_{CL} = 1$	Typ (V/μs)	3.5	0.3	0.3	0.5	1.5	0.5	0.5	0.5
$A_{CL} = -1$		3.5	0.3	0.3	6.0	2.5	6.0	6.0	6.0
$A_{CL} = 10$		5.0	3.0	3.0	2.0	8.0	2.0	2.0	2.0
Input Voltage Range	Max (V)	+1.5, −6.0	±10	±10	±15	±15	±15	±15	±15
Differential Input Voltage	Max (V)	±5.0	±5.0	±5.0	±30	±5.0	±30	±30	±30
Input Offset Voltage Drift	Typ (μV/°C)	10	1.8	3.0	7.0	3.0		3.0	3.0
	Max (μV/°C)						15		15
Offset Adjust					X		X	X	X
Output Short Circuit Protection					X		X	X	X
Compensated									
Dual						X			

* $V_S = +12, -6.0$ V
$V_S = \pm 15$ V, $T_A = 25°C$ unless otherwise specified

264

			General Purpose					High Power	Low Power
307	310	μA741	Compensated μA741E	μA776	μA747	1458		μA791	μA776
	Voltage Follower	Industry Standard	High Performance	Programmable $I_{SET} = 15\,\mu A$	Dual Industry Standard	Dual		1 Amp	150 μW Programmable $I_{SET} = 1.5\,\mu A$
7.5	7.5	6.0	3.0	6.0	6.0 ⟩	6.0		6.0	6.0
50	—	200	30	25	200	200		200	6.0
250	7.0	500	80	50	500	500		500	10
25	.999 × 10⁻³	20	50	50	20	20		20	50
±5.0	±5.0	±5.0	±5.0	±1.2	±5.0	±5.0		±5.0	±1.2
±18	±18	±18	±22	±18	±18	±18		±18	±18
1.0	20	1.0	1.0	1.0	1.0	1.0		0.2	0.2
0.5	30	0.5	0.7	0.7	0.5	0.5		0.5	0.1
0.5	—	0.5	0.7	0.7	0.5	0.5		1.0	0.1
0.5	—	0.5	0.7	0.7	0.5	0.5		6.0	0.1
±15	±10	±15	±15	±15	±15	±15		±15	±15
±30	—	±30	±30	±30	±30	±30		±30	±30
6.0	10	7.0	3.0	3.0	7.0	7.0		15	3.0
X	X	X	X	X	X			X	X
X	X	X	X	X	X	X		X	X
X	X	X	X	X	X	X			X
					X	X			

			General Purpose					High Power	Low Power
107	110	μA741	Compensated μA741A	μA776	μA747	1558		μA791	μA776
	Voltage Follower	Industry Standard	High Performance	Programmable $I_{SET} = 15\,\mu A$	Dual Industry Standard	Dual		1 Amp	150 μW Programmable $I_{SET} = 1.5\,\mu A$
2.0	4.0	5.0	3.0	5.0	5.0	5.0		5.0	5.0
10	—	200	30	15	200	200		200	3.0
75	3.0	500	80	50	500	500		500	7.5
50	.999 × 10⁻³	50	50	50	50	50		50	50
±5.0	±5.0	±5.0	±5.0	±1.2	±5.0	±5.0		±5.0	±1.2
±20	±18	±22	±22	±18	±22	±22		±22	±18
1.0	20	1.0	1.0	1.0	1.0	1.0		0.2	0.2
0.5	30	0.5	0.6	0.7	0.5	0.5		0.5	0.1
0.5	—	0.5	0.6	0.7	0.5	0.5		1.0	0.1
0.5	—	0.5	0.6	0.7	0.5	0.5		6.0	0.1
±15	±10	±15	±15	±15	±15	±15		±15	±15
±30	—	±30	±30	±30	±30	±30		±30	±30
3.0	6.0	7.0	5.0	3.0	7.0	7.0		10	3.0
15			15						
	X	X	X	X	X			X	X
X	X	X	X	X	X	X		X	X
X	X	X	X	X	X	X			X
					X	X			

Courtesy of Fairchild Semiconductor

SELECTION GUIDE
FOR HIGH ACCURACY
OPERATIONAL AMPLIFIERS

Selection Guide for Commercial Operational Amplifiers

		High Accuracy Instrumentation						
		Low Bias Current						
		FET μA740	Bipolar μA776	μA777	Super Beta 208	308	208A	308A
		High Z_{IN} $10^{12}\,\Omega$ High Slew Rate	Low Power $I_{SET}=1.5\,\mu$A					
Input Offset Voltage	Max (mV)	100	6.0	7.5	2.0	7.5	0.5	0.5
Input Offset Current	Max (nA)	0.3	6.0	50	0.2	1.0	0.2	1.0
Input Bias Current	Max (nA)	2.0	10	250	2.0	7.0	2.0	7.0
Voltage Gain	Min (V/mV)	25	50	25	50	15	80	80
Operating Supply Voltage Range	Min (V)	±5.0	±1.2	±5.0	±5.0	±5.0	±5.0	±5.0
	Max (V)	±22	±18	±20	±20	±18	±20	±20
Unity Gain Bandwidth	Typ (MHz)	3.0	0.2	1.0	1.0	1.0	1.0	1.0
Slew Rate $A_{CL}=1$	Typ (V/μs)	6.0	0.1	0.5	0.3	0.3	0.3	0.3
$A_{CL}=-1$		6.0	0.1	6.0	0.6	0.6	0.6	0.6
$A_{CL}=10$		6.0	0.1	2.0	—	—	—	—
Input Voltage Range	Max (V)	±15	±15	±15	±15	±15	±15	±15
Differential Input Voltage	Max (V)	±30	±30	±30	±0.5	±0.5	±0.5	±0.5
Input Offset Voltage Drift	Typ (μV/°C)	20	3.0	3.0	3.0	6.0	1.0	1.0
	Max (μV/°C)	—	—	—	15	30	5.0	5.0
Offset Adjust		X	X	X				
Output Short Circuit Protection		X	X	X	X	X	X	X
Compensated		X	X					

Selection Guide for Military Operational Amplifiers

		High Accuracy Instrumentation				
		Low Bias Current				
		FET μA740	Bipolar μA776	μA777	Super Beta 108	108A
		High Z_{IN} $10^{12}\,\Omega$ High Slew Rate	Low Power $I_{SET}=1.5\,\mu$A			
Input Offset Voltage	Max (mV)	20	5.0	2.0	2.0	0.5
Input Offset Current	Max (nA)	0.15	3.0	10	0.2	0.2
Input Bias Current	Max (nA)	0.2	7.5	75	2.0	2.0
Voltage Gain	Min (V/mV)	50	50	50	50	80
Operating Supply Voltage Range	Min (V)	±5.0	±1.2	±5.0	±5.0	±5.0
	Max (V)	±22	±18	±20	±20	±20
Unity Gain Bandwidth	Typ (MHz)	3.0	0.2	1.0	1.0	1.0
Slew Rate $A_{CL}=1$	Typ (V/μs)	6.0	0.1	0.5	0.3	0.3
$A_{CL}=-1$		6.0	0.1	6.0	0.6	0.6
$A_{CL}=10$		6.0	0.1	2.0	—	—
Input Voltage Range	Max (V)	±15	±15	±15	±15	±15
Differential Input Voltage	Max (V)	±30	±30	±30	±0.5	±0.5
Input Offset Voltage Drift	Type (μV/°C)	20	3.0	3.0	3.0	1.0
	Max (μV/°C)	—	—	15	15	5.0
Offset Adjust		X	X	X		
Output Short Circuit Protection		X	X	X	X	X
Compensated		X	X			